Environmental Science and Engineering

Environmental Science

Series Editors

Ulrich Förstner, Technical University of Hamburg-Harburg, Hamburg, Germany
Wim H. Rulkens, Department of Environmental Technology, Wageningen,
The Netherlands
Wim Salomons, Institute for Environmental Studies, University of Amsterdam,
Haren, The Netherlands

The protection of our environment is one of the most important challenges facing today's society. At the focus of efforts to solve environmental problems are strategies to determine the actual damage, to manage problems in a viable manner, and to provide technical protection. Similar to the companion subseries Environmental Engineering, Environmental Science reports the newest results of research. The subjects covered include: air pollution; water and soil pollution; renaturation of rivers; lakes and wet areas; biological ecological; and geochemical evaluation of larger regions undergoing rehabilitation; avoidance of environmental damage. The newest research results are presented in concise presentations written in easy to understand language, ready to be put into practice.

More information about this series at http://www.springer.com/series/3234

Kavindra Kumar Kesari
Editor

Networking of Mutagens in Environmental Toxicology

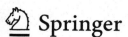 Springer

Editor
Kavindra Kumar Kesari
Department of Applied Physics, Department
of Bioproducts and Biosystems
Aalto University
Espoo, Finland

ISSN 1863-5520   ISSN 1863-5539   (electronic)
Environmental Science and Engineering
ISSN 1431-6250   ISSN 2661-8222   (electronic)
Environmental Science
ISBN 978-3-319-96510-9   ISBN 978-3-319-96511-6   (eBook)
https://doi.org/10.1007/978-3-319-96511-6

This Springer imprint is published by the registered company Springer Nature Switzerland AG
The registered company address is: Gewerbestrasse 11, 6330 Cham, Switzerland

Foreword

Networking of Mutagens in Environmental Toxicology is a welcome addition to the Springer Nature of Environmental Science and Engineering Series. In this book, the contributed authors have approached their topic in a truly multidisciplinary manner, where environmental toxicology plays a prominent role in their analysis by providing sufficient evidences on mutagenic factors. Pollutants are everywhere, and we are surrounded by it. They can be found in our surroundings that we are getting radiation exposures, the water that we drink, the air that we breathe, and the wastewater-irrigated crops that we eat. This book updates the theory and principle of environmental mutagens with their affecting factors like chemical, physical, and biological contaminations. Subjects covered in this book are mainly anthropogenic and natural pollution as well as feedback mechanisms and multiple stress or response to variable factors. Additionally, the effect of metal toxicity, ecotoxicity, nanotoxicity, and man-made radiations on human health has been discussed greatly. There are totally ten chapters, and these chapters mainly focus on the mutagenic factors and occurrence of mutation due to these in the environment. Discussions on the mechanisms and protective measures against mutagenic factors are the quality inclusion in this book. Toxic metal contaminations, and radiation induced toxicity may be one of the leading causes of concern for human health. This may induce oxidative stress and lead to neurodegenerative and infertility outcome discussed well in this book.

In recent years, many books have been published on the environmental toxicology. Mostly, they have specialized the chapters written by the contributing authors with specific area of research. Therefore, the special feature of this book is that the chapters are not repeated with a similar theme, where each and every chapter has multidisciplinary approach on the networking of mutagens in environmental toxicology. This shows great interest for readers where they can read and make an overview on existing mechanisms of toxicity in the environment. This book breaks the previous trends of specified chapters and offers a broad and coherent vision of the field. We must congratulate Dr. Kavindra Kesari (Editor) for the editing and to undertake very important topical issue in this book. This book has been designed with the aim for use as a reference book in courses offered to undergraduate, postgraduate, research scientists, and environmentalists. This book is also interesting for readers with non-scientific background. This book provides recent topics for individuals interested in the field of toxicology. We sincerely hope that the information embodied in this book will enthuse environmentalists and ameliorate upcoming new information.

Espoo, Finland   Prof. Tapani Vuorinen
                                                  Vice Dean, School of Chemical Engineering
                                                    Department of Bioproducts and Biosystems
                                                                            Aalto University

                                                                      Prof. Janne Ruokolainen
                                                          Director, Nanomicroscopy Center
                                                            Department of Applied Physics
                                                                            Aalto University

Preface and Acknowledgements

Networking of Mutagens in Environmental Toxicology updates the theory and principle of environmental mutagens by providing existing mechanisms of chemically, physically, and biologically interacting contaminants. Networking of mutagens is all around us and poses a problem in the onset of various diseases or physiological disorders to human beings due to man-made substances and compounds. Mutations occur because mutagens are mostly responsible for severe diseases like cancer, infertility, and neurodegenerative diseases. Various mutagenic factors, which pollute air, water, and food, possibly induce mutations in humans and are suspected of causing cancer. Such mutagens, which are responsible for causing mutations, are mainly natural or man-made radiations, heavy metals, stress, junk food, tobacco smoke, fine particles or nanoparticles, toxic chemicals, bacteria, viruses which have been discussed in this book. The networking of all these toxicants in the environment is responsible for toxicity and severe health issues. This book examines the mechanism for the environmental causes of cancer, classification of mutagens, metabolism of chemical and physical mutagens, and DNA damage and repair system. The chapters decipher the phenomena of defensive role of antioxidants and highlight the latest developments in environmental toxicology. The chapters include all basic to advance level of phenomena in solving the existing mechanisms of environmental toxicity. This book also highlights the latest developments in terrestial and aquatic ecosystem links to ecotoxicological effects. This book provides a sketchy outline about environmental toxicity, mutagenic factors, and associated health implications along with climatic and ecosystem balancing. This is not only exploring the theoretical outcomes but also giving an overview on experimental aspects. There are totally ten chapters, and all are connected and concerned with each other, where arisen problems have been opted and discussed by giving possible solutions. With multidisciplinary approach, this book extends a significant contribution by exploring possible measures in environmental toxicology and prevention.

Chapter "Role of Radiation in DNA Damage and Radiation Induced Cancer" mainly focuses to explore the overall aspects of ionizing and non-ionizing radiations. The study deciphers the fundamentals and the role of radiation exposures in DNA damage and related mechanisms associated with cancer. The man-made radiations are one of the high causes of concerns in the twenty-first century, where these radiations have causative effects on brain and associated organs. Moreover, the heavy metal contamination has also been identified as one of the growing health concerns. Therefore, Chapter "Mechanistic Effect of Heavy Metals in Neurological Disorder and Brain Cancer" illustrates the metal evoked mechanism, which impaired the function of neurons and generated the neurotoxicity and neurodegenerative diseases. Toxic metals like cadmium (Cd), lead (Pb), arsenic (As), mercury (Hg), thallium (Th)-induced oxidative stress may initiate the signaling processes, which activates the glial cells to produce immunogenic response and leads to the neuronal cell death. Heavy metals or heavy metal-containing compounds have been identified to be potent mutagens and carcinogens. Therefore, Chapter "Molecular Mechanisms of Heavy Metal Toxicity in Cancer Progression" has provided an additional mechanism to understand the phenomena of cancer formation due to heavy metal-induced toxicity. In this chapter, the authors mainly used molecular pathway analysis to understand the toxicity and carcinogenicity of heavy metals. Because, heavy metals may induce oxidative stress, and cause DNA damage, cell death and several signaling pathways which may leads to an increased risk of cancer and cancer-related diseases. In this series, Chapter "Burden of Occupational and Environmental Hazards of Cancer" represents the existing environmental factors, which are in our surroundings. The chemical, metals, radiations, smoke, asbestos, etc., are major kinds of environmental pollutants causing serious health issues. These pollutants have been classified into cancer-causing elements by International Agency for Research on Cancer. This chapter explores both causing factors and preventive measures for cancer. In this connection, Chapter "Environmental Toxicants and Male Reproductive Toxicity: Oxidation-Reduction Potential as a New Marker of Oxidative Stress in Infertile Men" shows causative effects of several lifestyle factors including heavy metals or trace elements, which have been found responsible to cause infertility. Expoure to various environmental factors may induce oxidative stress and discrete the measures of reactive oxygen species, total antioxidant capacity, and post-hoc damage suggest an ambiguous relationship between the redox system and male fertility. Such mutagenic oxidative stress can be defined by system biology, which has great applications in medical research. Therefore, Chapter "System Network Biology Approaches in Exploring of Mechanism Behind Mutagenesis" deciphers the fundamentals for the determination of different properties of the networks that help in analyzing the network graph and finding the most probable network that best describes the process. System network biology helps in providing new perspectives of inspection of these biological systems in the form of networks with the help of mathematical representations. Therefore, this study mainly investigates the candidate protein molecule that may act as a target protein with the help of network analysis. Chapter "Ecotoxicological Effects of Heavy Metal Pollution on Economically Important Terrestrial Insects" reveals biological responses toward heavy metal pollution in living organisms like insects. However, insects play a crucial ecological role in the maintenance of ecosystem structure and functioning. Therefore, any kind of metal or other mutagenic contaminations may kill the natural predators, which may disturb the ecosystem. It is quite clear now that our environment is being highly polluted due to man-made sources and there are a number of ways which may affect the human health by toxic contaminants such as metal, radiation, bacteria through existing sources like air, water, soil. In this series, Chapter "Contamination Links Between Terrestrial and Aquatic Ecosystems: The Neonicotinoid Case" represents the existence of neonicotinoid in environmental toxicity. Regarding pesticide pollution, the case of neonicotinoid insecticides is one of the most pressing issues nowadays, which classifies a family of toxic substances based on its known factor as clear stressors of natural ecosystems. Therefore, this chapter mainly aimed to review the current state of knowledge about neonicotinoid contamination in terrestrial and aquatic ecosystems and the toxic effects of NNIs to nontarget organisms, living in these ecosystems. An increasing human population is one of the leading mutagenic factors on this ecosystem, where rare diseases are being more generalized in the form of aging or neurodegenerative diseases. Therefore, Chapter "RAGE Exacerbate Amyloid Beta (Aβ) Induced Alzheimer Pathology: A Systemic Overview" is to demonstrate the evaluation of thoughts regarding the Alzheimer's disease pathway and the present information on the glyoxalase enzymes and their documented role in the control of glycation processes. In this chapter, an important role of biomolecular glycation-induced AGE and its interaction with expressed RAGEs are surveyed which shows new treatment strategies for age-related complications, and AD shows high possibilities for future research. RAGE plays a decisive role in the pathogenesis of AD, promoting the aging and related disorders by causing synaptic and neuronal circuit dysfunctions and Aβ-tau phosphorylation, which has been discussed in this chapter. Chapter "Elucidation of Scavenging Properties of Nanoparticles in the Prevention of Carcinogenicity Induced by Cigarette Smoke Carcinogens: An In Silico Study" deciphers in silico approach to find molecular interaction of cigarette smoke carcinogens with enzymes involved in DNA repair pathways. It shows scavenging capacities of the nanoparticles for polycyclic aromatic hydrocarbons (PAHs) and other toxicants by measuring the interaction of enzymes, proteins, and DNA repair pathways. In consideration of the future possibilities, it could be a problem-solving tool of experimental research. The medicinal use of nanoparticles significantly grows since a decade. I hope this book will serve as both an excellent review and a valuable reference for formulating suitable measures against environmental toxicology and for promoting the science involved in this area of research.

Finally, I would like to dedicate this book to my mother, late Parwati Devi, for her blessings. I am also highly thankful to my father, Dr. Arjundas Kesari, who has given me much inspiration and support. I would also like to thank Prof. Syed Waseem Akhtar, Founder and Chancellor (Integral University, Lucknow, India), who encouraged me for this initiative. I would like to thank all the authors who have contributed to this book. Finally yet importantly, my special thanks to series editor, publisher, and entire Springer team for their sincere assistance and support.

Espoo, Finland   Kavindra Kumar Kesari, Ph.D.

Contents

Role of Radiation in DNA Damage and Radiation Induced Cancer

Vaishali Chandel, Gaurav Seth, Priyank Shukla and Dhruv Kumar

Abstract Radiation has been reported to be a proven carcinogen which is responsible for more than half of all malignancies. The incident rates, morbidity and mortality of these cancers are increasing and thus reflects a serious health concern in public. Ionizing and Non-ionizing radiation exposure both lead to the development of cancer. Ultraviolet radiation (UVR) which is a non-ionizing radiation damages the DNA and causes genetic mutations. Exposure to Ionizing radiation results in the various oxidizing events altering the structure of atoms through direct interactions of the radiation with the target molecules or via the product of radiolysis of water. In this chapter we have discussed about the role of radiation in DNA damage and related mechanisms associated with cancer.

Keywords Radiation · DNA damage · Mutation · Cancer

1 Introduction

Radiation can be broadly classified as Ionizing radiation (IR) and non-ionizing radiation (NIR). Ionizing radiation has shorter wavelengths with higher frequencies and at the molecular level, has sufficient energy to generate ions in matter by breaking the chemical bonds. It is emitted as very high energy electromagnetic waves such as x-rays and gamma rays from the radioactive atomic structures. Non-ionizing radiation, on the other hand, has long wavelength with low frequency. In the electromagnetic spectrum, the types of non-ionizing radiation include ultraviolet (UV) rays, visible spectrum, infrared rays, radio frequencies, microwave frequencies (Ramasamy et al. 2017).

V. Chandel · G. Seth · D. Kumar (✉)
Amity Institute of Molecular Medicine and Stem Cell Research (AIMMSCR), Amity University Uttar Pradesh, Sec-125, Noida, India
e-mail: dhruvbhu@gmail.com; dkumar13@amity.edu

P. Shukla
Northern Ireland Centre for Stratified Medicine, Biomedical Sciences Research Institute, Ulster University, C-TRIC Building, Altnagelvin Area Hospital, Glenshane Road, BT47 6SB Derry/Londonderry, Northern Ireland, UK

© Springer Nature Switzerland AG 2019
K. K. Kesari (ed.), *Networking of Mutagens in Environmental Toxicology*, Environmental Science,
https://doi.org/10.1007/978-3-319-96511-6_1

2   Ionizing Radiation

When the living cells absorb ionizing radiation, it leads to the disruption of atomic structures which ultimately results in the production of biological as well as chemical changes. Ionizing radiation also acts in an indirect manner through water radiolysis, and therefore results in the generation of reactive species that would damage nucleic acids, proteins and lipids (Arvelo et al. 2016). These indirect and direct effects of radiation together initiate a cascade of molecular and biochemical signalling events that can cause permanent changes in the cell physiologically or may result in cell death (Spitz et al. 2004).

2.1   Primary Effect of Ionizing Radiation

2.1.1   Radiolysis of Water and Reactive Oxygen Species Generation

When water absorbs the energetic radiations, it leads to excitations and ionizations resulting in the free radicals generation which attack other critical molecules of the cell. This complex cascade of events can be classified into four stages. The first stage known as the physical stage involves the incident radiation causing the deposition of energy and generation of secondary electrons. During the second stage, also known as the physiochemical stage, these electron species are highly unstable and undergo reorganization resulting in the production of molecular as well as radical products of radiolysis. The third stage involves the diffusion and reacting of the different chemical species with each other and also with the environment. In the last stage known as the biological stage involves the responding of the cells to damage caused from the generated products from the previous three stages. This stage basically involves the biological responses to cells induced by radiation exposure (Azzam et al. 2012).

2.1.2   Reactive Nitrogen Species Generation

The generation of high amount of Nitric oxide ($\cdot$NO) is mediated by ionizing radiation that stimulates Nitric oxide synthase activity in the cells. Similar to the $\cdot$OH radical, Peroxynitrite (ONOO−) is extremely capable of attacking and damaging the different targets of cell including thiols, DNA bases, proteins, lipids since they are highly reactive (Azzam et al. 2012).

Thus, the generation of the reactive species induced by radiation has severe consequences onto the cell leading to DNA damage. Reactive nitrogen and oxygen species can attack and damage DNA by causing base damage, sugar crosslinks, DNA breaks, telomere dysfunction, and destruction of sugars leading to alterations in DNA (Sahin et al. 2011; O'Neill and Wardman 2009; Valerie et al. 2007). If they are not repaired, these can result in the mutations causing cell death or promote transformation of the cell making it carcinogenic (Kryston et al. 2011).

2.1.3 Effect of Ionizing Radiation on Mitochondria

When the cell is exposed to the ionizing radiation, the metabolic activity of the cell is hampered and may lead to the disruption of mitochondria function because of various reasons such as antioxidants modulation, alteration of metabolic activity in response to oxidative damage, ROS generation (Leach et al. 2001).

Mitochondria is known to be the powerhouse of the cell via aerobic respiration involving the Electron transport chain, Kreb's cycle and oxidative phosphorylation pathway. It has been reported that mitochondria plays a major role in the generation of reactive oxygen species by consuming 85–90% of the body's oxygen (Cadenas and Davies 2000; Babior 1999; Los et al. 1995; Alberto and Chance 1973). About 1–5% of the electrons generated in the respiratory chain of mitochondria is diverted for the generation of superoxide radicals mediated by ubiquinone-dependent reduction mechanism (Boveris et al. 1976). Therefore if the mitochondria is disrupted or is dysfunctional in cells induced by radiation results in the disturbance in the oxidation-reduction reactions.

During the oxidative phosphorylation process, from complex 1 and complex 3 (complex 2 to a minor extent) if there is an electron leakage prematurely, than it leads in the reduction of oxygen (O_2) molecule to generate superoxide (O_2) (Azzam et al. 2012). Besides this, the induced radiation leads to further excess leakage of the electrons from the electron transport chain, and thus resulting in the excess generation of oxygen radical O_2 (Droge 2002). The production of excess—(ROS)—may lead to mutation in the mitochondrial DNA and alteration in the expression of protein and cell damage which are essential for the cellular and mitochondrial functions.

The Reactive oxygen species (ROS) generation or Reactive Nitrogen species (RNS) generation by the induced radiation at the early and late stages leads to changes in copy number of DNA (Malakhova and Bezlepkin 2005), genomic instability, mutations in DNA, altered gene expression (Chen et al. 2003; Chaudhry and Omaruddin 2011), autophagy (Lomonaco et al. 2009; Chiu et al. 2011), DNA damage (Choi et al. 2007), and transformation of cells (Du et al. 2009).

2.2 Protein Import in Mitochondria

Mitochondria are known to contain its own DNA and the entire system for replication, transcription and translation. However, mitochondria synthesizes only certain (13 in humans) proteins (Azzam et al. 2012). The rest of the proteins of mitochondria are encoded by nucleus and are synthesized by cytoplasmic ribosomes. It is important that the proteins are significantly transported to the sub compartments of mitochondria from the cytoplasm correctly (Pain et al. 1990). Hence for the mitochondrial biogenesis the fundamental principle involves the import of protein into mitochondria. If there is a defect in the protein import in mitochondria caused by induced ionizing radiation, it may result in the amplification of oxidative stress and can cause certain health effects including metabolic disorders and degenerative diseases (Azzam et al. 2012).

3   Non-ionizing Radiation

Ultraviolet radiation (UVR) has been reported to be a major carcinogen responsible for majority of the skin cancer. Depending upon the wavelength, UVR is categorized into three types such as ultraviolet C (UVC; 100–280 nm) having the shortest wavelength and highest energy, ultraviolet B (UVB; 280–320 nm), and ultraviolet A (UVA; 320–400 nm) with the longest wavelength and least energy (Orazio et al. 2013). The penetration of UV into the skin is in the wavelength dependent manner. Among the different types of UVR, the mutagenic rate of UVC is very high but it does not reach the surface of the earth because it is completely absorbed by the ozone layer of the stratosphere (Ramasamy et al. 2017).

3.1   Responses to UV

3.1.1   Direct Damage

When the DNA absorbs UV rays, it causes the molecular rearrangements leading to the formation of specific cyclobutane pyrimidine dimers (CPDs) and pyrimidine (6-4) pyrimidonephotoproducts (6-4PP) which are highly mutagenic. The damage caused by UV rays are mainly responsible for skin cancer because of the mutation caused in DNA (Mallet et al. 2016). UVA having the maximum penetration power has the potential to generate Reactive oxygen species (ROS) which via the indirect photosensitizing reactions can cause DNA damage. The generation of ROS and DNA lesions for example cyclobutane pyrimidine dimers (CPDs) and pyrimidine (6–4) pyrimidone photoproducts (6-4PP) is mediated by UVB that can lead to primary as well as secondary breaks in DNA. These DNA lesions are responsible for halting transcription/replication leading to the production of DNA double—strand breaks (DSBs) near stalled replication fork at CPDs—containing DNA (Batista et al. 2009; Limoli et al. 2002). These DNA lesions leads to the generation of oxidative product, strand breaks which if not repaired results in mutation in DNA causing tumourigenesis and damage to cell by disruption of fundamental processes (Takahashi and Ohnishi 2005) (Table 1).

3.1.2 Indirect Damage

Indirect damage to DNA in response to Ultraviolet radiation occurs due to the generation of ROS which is majorly caused by UVA. It has already been reported that UVA irradiation is the leading cause and plays a major role in skin carcinogenesis and photoaging by ROS generation such as superoxide anions, hydrogen peroxide, and hydroxyl radical (Takeuchi et al. 1998; Meyskens et al. 2001).

When a chromospheres excited by light does not return to it's original ground state, a phenomena known as photosensitization occurs by generation of heat or emission of photons and leads to the initiation of chemical reactions which ultimately leads to the generation of reactive species. Therefore the physical nature of UVA and the interaction with absorbing chromophore in the skin generates ROSin the skin. This ROS generation in response to UVA acts as a powerful mutagen that causes oxidative DNA damage (Wondrak et al. 2006) (Table 1).

Since the Nucleotides are very much prone to free radical injury. If there is an oxidation of nucleotide bases, it promotes mispairing among the bases causing mutation (Arvelo et al. 2016). When the DNA is exposed to UVA radiation, a number of oxidative products of pyrimidine bases such as 6-hydroxy-5,6-dihydro-5-yl, -5-hydroperoxy-6-hydroxy-5, 6-dihydropyrimidine and purine bases such as 8-oxo-Ade,2,6-diamino-4-hydroxy-5-formamidoguanine(FapyGua),8-oxo-7,8-dihydroguanyl(8-oxoGua), oxazolone and FapyAde have been reported to form (Ravanat et al. 2001).

3.2 Ultraviolet Radiation Exposure Induces Carcinogenesis

As reported skin cancers are the most common type malignancies in humans, having over a million cases diagnosed each year (Rogers et al. 2010). Skin cancers can be categorized as melanoma and non-melanoma skincancers (NMSC), depending on the origin of cell and behaviour. The exposure to Ultraviolet radiation is the leading cause of skin cancer (Narayanan et al. 2010). The most dangerous form of skin cancer is Malignant melanoma arising from the epidermal melanocytes (Berwick and Wiggins 2006). Early detection of melanomas can be easily treated by surgical excision alone. However, melanomas are quick to invade and metastasize and long-term survival is poor for advanced disease. There has lately been progression in terms of targeted therapy and immunotherapy, but however melanoma are most difficult to treat once it has already been migrated from its original site (Orazio et al. 2013). Non-melanoma skin cancers are less fatal compared to melanoma and can be treated easily because of the tendency to remain confined to their original primary site of disease and also has the long term prognosis (Orazio et al. 2013). There are strong epidemiologic as well as the molecular data suggesting that there is a direct connection of skin cancer to UV exposure (Poulton et al. 2013) and it has been roughly estimated that the causative agent for approx 65% of melanoma and 90% of non-melanoma skin cancer is UV radiation (Armstrong and Kricker 1993; Pleasance et al. 2010).

The process of development of skin carcinogenesis mediated by UVR is a three step process including initiation, promotion and progression which is mediated by different cellular, biochemical, and molecular changes (Arvelo et al. 2016). During the process of initiation the normal keratinocytes gain the ability to form tumours which is an irreversible process. While during the process of promotion which is basically reversible, the keratinocytes which have formed clone expand to develop into a papilloma (Arvelo et al. 2016). Tumour progression involves a series of genetic as well as the epigenetic events transforming the premalignant papilloma into malignant squamous cell carcinoma. The cells become resistant to the signal transduction pathways for terminal differentiation which is because of the genetic alteration in proto-oncogenes and tumour suppressor genes. There is an overexpression and mutation in tumour suppressor gene p53 in the interfollicular epidermis due to the constant ultraviolet radiation exposure leading to squamous cell carcinoma. This mutation is beneficial for the cell growth and prevents the cell from undergoing apoptosis (Rebel et al. 2012).

3.3   Pigmentation

The most abundant cells in the layer of epidermis are Keratinocytes and accumulate melanin pigments as they mature, and the function of epidermal melanin is to block the penetration of ultraviolet radiation into the skin (Slominski 2004). There are two major chemical forms of melanin; (i) eumelanin, a dark pigment expressed in very high amount in the skin of individuals who are heavily pigmented and (ii) pheomelanin, a sulphated pigment of light colour which is present in melanin precursors leads to the incorporation of cysteines (Ito et al. 2000). The eumelanin form is very efficient in blocking the ultraviolet radiation compared to pheomelanin, therefore if in the skin the amount of eumelanin is more, than it will be less permeable to the ultraviolet radiation (Vincensi et al. 1998). A scale known as the "Fitzpatrick scale" which is made up of six different phototypes is a semi-quantitative scale which describes the colour of the skin by determining the inflammatory response to UV, levels of melanin, basal complexion and cancer risk (Scherer and Kumar 2010). A quantitative method known as the Minimal Erythematous Dose (MED) reports the ultraviolet radiation amount (especially UVB), which is responsible to induce sunburn for 24–48 h in the skin after the exposure by determining edema (swelling) and erythema (redness) as endpoints.

It has been observed that there is a high risk of skin carcinogenesis in the people who are fair-skinned because they are more sensitive to the ultraviolet radiation and have very less amount of epidermal eumelanin. The levels of pheomelanin is almost similar between the light-skinned and dark-skinned individuals and thus the levels of epidermal eumelanin plays a major role in determining the complexion of the skin, sensitivity of ultraviolet radiation and risk of cancer. However pheomelanin leads to the oxidative damage and causes DNA injury by the generation of free radicals in melanocytes even when there is an absence of ultraviolet radiation (Benedetto et al. 1982; Prota 2000; Sealy et al. 1982; Mitra et al. 2012).

In summary, it is easier for the ultraviolet radiation to cause inflammation in the fairer skin compared to the darker-skinned individuals. In dark-skinned individual, Minimal Erythematous Dose (MED) is highest because to burn eumelanin rich skin, more ultraviolet radiation is needed (Kawada 2000; Lu et al. 1996). And Since fair-skinned people have low Minimal erythematous dose as compared to dark-skinned, ultraviolet radiation can penetrate easily and damage the skin (Ravnbak 2010).

4 Genes Involved in DNA Damage

Ionizing radiation has proven to be one of the important carcinogen since it leads to the generation a bulk of injuries in DNA of the human cells including single and double stranded breaks (DSBs) and various base damages (Chistiakov et al. 2008). Double stranded breaks (DSBs) are lethal and deadliest form of DNA damage, since an unrepaired DSB can lead to cell death (Hein et al. 2014). Various cellular DNA repair pathways are responsible for correction of radiation induced DNA damage. Double stranded breaks (DSBs) involve two major mechanisms, homologous recombination (HR) and nonhomologous end joining (NHEJ) (Agarwal et al. 2006).

4.1 Genes Involved in Homologues Recombination Pathway of DNA Repair

4.1.1 Ataxia Telangiectasia Mutated (ATM)

The ATM gene encodes for an important checkpoint kinase in cell cycle and regulates a variety of downstream proteins, including BRCA1, p53 which is a tumour suppressor, various DNA repair proteins such as NBS1 and SMC1, the proteins involved in checkpoint RAD9 and RAD 17, and checkpoint kinase CHK2. SMC1 plays an important role in and DNA repair after the damage has occurred and controlling DNA replication forks. ATM phosphorylates and activates SMC1. BRCA1 and NBS1 recruit activated ATM to DNA break sites. ATM then phosphorylates SMC1 (Chistiakov et al. 2008). The cells which are ATM-deficient are highly sensitive to ionizing radiation induced DNA damage (Morrison 2000). However the ATM- deficient cells have the capability to repair most of the DSBs induced by radiation with the normal kinetics, but they fail to repair a subset of breaks irrespective of the initial number of lesions induced. Also, as compared to the NHEJ-defective cells, after irradiation these ATM-deficient cells do not have the ability to recover. These observations suggest the extreme sensitivity of patients with ATM to ionizing radiation (Morrison 2000).

The mutations in ATM gene leads to an autosomal recessive disorder known as Ataxia telangiectasia. It has been observed that Ataxia telangiectasia patients are more prone to develop cancer. More specifically, obligate heterozygous carriers of mutations in ATM may have an increased risk of developing breast cancer (Chistiakov et al. 2008).

4.1.2 RAD51

The gene family of RAD51 involves various proteins which show ATPase activity stimulated by DNA and typically for single-stranded DNA binding and forming complexes with each other (Thacker 2005). In DNA damage response pathway, RAD51 is involved and leads to the activation of DSB and HR repair. RAD51 shows DNA-dependent ATPase activity and has the ability to bind to the single and double stranded DNA. RAD51 unwinds the duplex DNA and at the site of DNA break forms helical nucleoprotein filaments (Chistiakov et al. 2008). In the promoter region of RAD51 gene, two SNPs (-135 G>C) and (-172 C>T) have been found which shows both are functional and increases the RAD51 promoter activity (Hasselbach et al. 2005). In the patients of breast cancer majorly at the subgroup involving the mutations in BRCA2 (Kadouri et al. 2004; Levy-Lahad et al. 2001) and acute myeloid leukaemia (Seedhouse et al. 2004) shows a strong association with 135 G>C SNP. It was reported that there is a significant association of $+135$C allele of RAD51 with raised risk of radiotherapy-induced acute myeloid leukaemia (Jawad et al. 2012). The G to C substitution of RAD51 at 135 position is a gain-of-function mutation, which leads to an increased RAD51 activity. In addition, increased RAD51 levels have been observed in different tumour lines (Fan et al. 2004) suggesting that the increased levels of RAD51 recombinase may play a major role in increased risk of tumourigenesis process (Raderschall et al. 2002).

4.1.3 BRCA1 and BRCA2

BRCA1 plays a major role in various cellular processes such as regulation of the cell cycle, chromatin remodelling, transcriptional regulation. In the early steps of DNA repair, BRCA1 plays an important role in the promotion and regulation of HR. The kinases such as RAD3 related, ATM, checkpoint kinase 2 phosphorylates BRCA1, in response to the Double stranded breaks, and plays an important role in signal transduction induced by DNA damage (Chistiakov et al. 2008). The function of BRCA2, on the other hand, is confined to Recombination and DNA repair only. The core HR mechanism is regulated by BRCA1 and BRCA2 via RAD51 recombinase control since BRCA1 and BRCA2 binds to RAD51 via BRC repeats which are eight evolutionary conserved binding domains (Pellegrini et al. 2002). Mutations in the genes of BRCA1 and BRCA2 are strongly associated with the development of breast and ovarian cancer. Heterozygous women for mutations in BRCA1 or BRCA2 have a higher chance to develop breast and ovarian cancer (Honrado et al. 2005). Ionizing radiation induced Double stranded breaks repair in human carcinoma cells lacking BRCA1 and BRCA2 gene suggests the rejoining of DNA DSBs normally. This shows that there may not be a direct role of BRCA1 or BRCA2 in the rejoining of DSBs induced by radiation in the genome of tumour cells (Honrado et al. 2005).

4.1.4 MRN Complex

MRN complex plays a major role in the detection of DNA damage and DNA damage response activation. MRN complex is formed by three major proteins RAD50, MRE11 and NBS1. When the DNA is damaged, the MRN complex binds to the broken ends of DNA and undergoes a series of conformational changes that lead to the activation of ATM (Paull and Lee 2005).

MRE11 mutations results in an autosomal recessive disease Ataxia-telangiectasia like-disorder (ATLD) characterized by a raised sensitivity to the exposure of radiation (Chistiakov et al. 2008).

4.2 Genes Involved in the Non-homologous End Joining Pathway of DNA Repair

4.2.1 DNA-Dependent Protein Kinase

A multiprotein complex known as DNA-dependent protein kinase (DNA-PK) has the catalytic and regulatory subunit. The catalytic subunit consists of DNA protein kinase; DNA-PKcs and the regulatory subunit consist of Ku heterodimer. The catalytic subunit binds to the Ku heterodimer leading to the formation of DNA-PK complex that triggers DNA-PKcs activity through autophosphorylation. Ku heterodimer which is comprised of two subunits Ku70 and Ku80 binds to the free ends of the DNA at the break to facilitate them in close proximity (Chan and Chen 2002). Phosphatidylinositol-3-kinase which is a serine/threonine kinase family has the member DNA-dependent protein kinase (DNA-PK) as part. When DNA-PK is phosphorylated, it becomes active and phosphorylates various other proteins participating in DNA repair such as XRCC4, Artemis, Ku70, Ku80, H2AX and others (Collis et al. 2005).

4.2.2 XRCC4/DNA Ligase IV Complex

The core components of NHEJ complex involves DNA-PK, DNA ligase IV and XRCC4 (Chistiakov et al. 2008). Even if the antiparallel strand cannot be ligated, XRCC4/DNA ligase IV has the capability to ligate one strand and thus the remaining SSB can be repaired as a single-stranded lesion. It has been reported that the radio sensitivity and genomic instability is promoted if there is a genetic alterations in XRCC4 and DNA ligase IV. Mutation in XRCC4 leads to the deficiency of radiosensitive phenotype and protein of respective cell lines (Chistiakov et al. 2008).

4.2.3   Artemis Nuclease

Artemis is a nuclease which has the endonuclease activity at $5'-3'$ that removes overhangs at $5'$ and shortens $3'$. It has been observed that when DNA-PK phosphorylates Artemis, it activates the Artemis hairpin activity which is essential for V(D)J recombination (Ma et al. 2002). It was found that after exposure to ionizing radiation, Artemis is a target for DNA-PK (Soubeyrand et al. 2006) and ATM dependent (Zhang et al. 2004) mediated phosphorylation.

4.2.4   DNA Polymerase μ

DNA polymerase is an important enzyme in NHEJ facilitating the successful repair of DSBs. It has been shown that with the interaction of Ku, DNA polymerase μ plays a major role in radiation-induced DNA repair dependent on the XRCC4/ligase IV complex. Although till date there is no report of an association of *polymerase μ* with radiation-induced cancer (Chistiakov et al. 2008).

5   Epigenetic Regulation

Epigenetic modifications are heritable changes in the function and structure of the genome that happens without the change in the DNA sequence. The post-translational alterations in histones and DNA methylation are the common epigenetic changes observed in mammalian cells (Kim et al. 2013). These modifications at the epigenetic level have been reported to play a critical and important biological role in multiple organisms such as the normal growth, development, and differentiation (Kim et al. 2013).

It has been reported that the abnormal mechanism of epigenetic modulation displays the global changes including the chromatin packaging as well as in localized promoter changes in the gene, which influence the gene transcription involved in the cancer development (Peter and Baylin 2010) (Table 1).

5.1   Radiation and DNA Methylation

DNA methylation is the most common mechanism responsible for the epigenetic regulation. It includes majorly two types of alteration; (i) hypomethylation and (ii) hypermethylation of certain genes and the global genome DNA methylation changes (Kim et al. 2013). Reports suggests that the exposure to Ionizing radiation could hamper the patterns of DNA methylation. IR exposure has been observed to be dose dependent (Pogribny et al. 2004).

Loss of methylation patterns, hypomethylation coupled with the decrease in the expression levels of DNA methyl CpG binding proteins (MeCP2) and methyltransferases (DNMTs; DNMT1, DNMY3a, and DNMT3b) and was linked with the changes induced by radiation (Raiche et al. 2004; Loree et al. 2006). These results suggest that in the exposed animals due to radiation exposure, DNA hypomethylation patterns could lead to genomic instability (Kim et al. 2013).

5.2 Histone Modification

The modulation of normal epigenome considering in aspect of maintaining the pattern of gene expressions and the normal structure of chromosome as well as function by nucleosome positioning with 147 bp of the DNA coiled around the core histone octamer (H2A, H2B, H3, and H4) (Kim et al. 2013; Jenuwein and Allis 2001). The modification of chromatin for packaging of DNA and DNA methylation strong dependence is well established. Besides this the modification of histones and DNA methylation closely interact in the initiation of chromatin transcriptional state. The expression of such genes with no methylation patterns shows that there promoters have identical distribution of the active marks H3K4me and H3K9acetyl (Nguyen et al. 2002; McGarvey et al. 2006).

In contrast, when the genes associated with silencing are associated with hypermethylation, the active markers distribution is drastically decreased and virtually every histone methylation mark such as monomethylation, dimethylation, and trimethylation of H3K27 and H3K9 which has been linked with the repression at the transcriptional level is enhanced (McGarvey et al. 2006).

The phosphorylation of γH2AX induced by radiation was significantly studied as a measure of Double stranded breaks (DSB) accumulation in the cells which are irradiated (Bonner et al. 2011). Regarding the Ionizing radiation exposure, histone phosphorylation at serine 139 (γH2AX) is studied well as the modification of histones affected by the exposure of IR since γH2AX is one of the first signals against DSB for a cellular response and IR exposure (Kim et al. 2013; Pilch et al. 2003; Rogakou et al. 1998).

γH2AX accumulates in the nucleus at the double stranded breaks leading to the formation in the γH2AX loci, and a direct interrelation has been observed between the phosphorylation of γH2AX and the number of DSBs occurring from radiation. Therefore, for the maintenance of the stability of genome as well as for the DSB repair, γH2AX is very important and crucial factor (Celeste et al. 2003).

6 Small RNAs

Small RNAs e.g. microRNA (miRNA) also controls the epigenetic regulation. MicroRNAs (miRNAs) are short, non-coding ssRNA, 18-25-nucleotide long, which plays an important role in the regulation of gene expression at post-transcriptional level via binding to their target protein which encodes for the respective mRNAs (Bartel 2004; Singh and Campbell 2013). Estimation suggest that 30% of the human genome is under miRNA regulation and more than 1000 miRNAs are transcribed. Also one miRNA can lead to the modulation of hundreds of down-steam genes post-transcriptionally (Kim et al. 2013). They control a diversed range of biological process such cellular proliferation, cellular differentiation, stem cell maintenance and apoptosis (Bernstein and Allis 2005). The loss of function in miRNAs can initiate and lead to the malignant transformation of normal cells and can make them act as tumour suppressors (Kim et al. 2013). Various mechanisms such as mutation, genomic deletion, epigenetic silencing and alteration in miRNA processing are responsible for the loss of function of a miRNA (Nakamura et al. 2007; Calin et al. 2004; Saito et al. 2006). miRNAs are also responsible in IR-induced response in vivo as well as in vitro. Studies on the effects of IR exposure on the whole body of rodent, revealed altered expression pattern in miRNA in tissue-specific as well as sex-specific protective mechanism (Ilnytskyy et al. 2008).

7 PARP—Poly (ADP-Ribose) Polymerase

Poly (ADP-ribose) polymerases (PARPs) comprises a large family including 18 proteins (Soldatenkov and Smulson 2000). During DNA damage, PARP1 and PARP2 enzymes are activated facilitating DNA repair in various pathways including the base excision repair (BER) and single strand breaks (SSBs) (Heale et al. 2006). PARP1 consists of three major domains which are conserved; (i) an auto modification domain, (ii) DNA damage sensing, and (iii) binding domain comprising of three zinc fingers at NH2 terminal. The strongest affinity for DNA breaks is mediated by Zinc finger 2, while Zinc finger 1 and 3 is responsible for the activation of DNA dependent PARP-1 (D'Arcangelo et al. 2016).

PARP1 is a chromatin-associate of NAD^+ covalently bonded to itself and various other nuclear acceptor proteins such as to the side chains of asparagine, arginine, glutamic acid, lysine, serine, and cysteine residues onto its substrates (Liu et al. 2017; Soldatenkov and Smulson 2000).

7.1 Role of PARP in DNA Repair Pathway

7.1.1 Base Excision Repair/SSB Repair Process

Many chemical modifications in cells such as deamination, methylation, oxidation and hydroxylation may lead to the induction of base damage and Single strand breaks (SSBs). In Base excision repair, DNA glycosylases cleave the damaged bases leading to the production of a basic sites which are further processed by AP nuclease (APE) into SSBs (Wei and Yu 2016). Later these damaged sites are repaired with the help of two different pathways known as short-patch repair and long-patch repair. PARP1 can functionally interact with SSRB factor XRCC1, facilitating the assembly and recruitment of SSRB machinery (Wei and Yu 2016). There are several reports suggesting that in the BER/SSBR process, PARP1 is able to interact with various factorssuch as XRCC1, DNA polymerase, DNA glycosylase 8-oxoguanine glycosylase 1 (OGG1) (DNAP) β, DNA ligase III, Aprataxin, proliferating cell nuclear antigen (PCNA) which can undergo PARylation by PARP1. In addition, PARP2 has been identified to interact with BER/SER proteins such as DNA ligase III. Additionally, PARP2 has also been identified to interact with BER/SSBR proteins such as DNAP β, and DNA ligase III, XRCC1. These information suggest that PAR chain plays a crucial role in the recruitment of DNA repair complexes (Wei and Yu 2016).

8 Prevention from Radiation

8.1 Role of T4N5 Liposome Lotion in Enzyme Therapy of Xeroderma Pigmentosum

Xeroderma Pigmentosum (XP) is a hereditary disease, caused by extreme exposure to the Sun leading to excessive sunburn. In the DNA repair mechanism induced by Ultraviolet radiation, and specifically in the repairing of CPDs, it has been observed that there is a biochemical defect in XP patients. In initiation of removal of Ultra violet radiation induced CPDs and in mammalian cells replacing of incision enzymes, T4 endonuclease V bacteriophage plays a major role. In mammalian cells this is expressed by the transfection of denVgene, which enhances the repairing process of CPDs and also reduces the frequency of Ultraviolet radiation induced mutations. T4N5 liposomes are generated by purified recombinant T4 endonuclease V in liposomes in a pH sensitive membrane. T4N5 liposomes in a hydrogel lotion, is applied to the skin for the prevention from Xeroderma Pigmentosum (Yarosh et al. 1996).

Table 1 Genetic and epigenetic changes associated with radiation induced cancer

Type of Radiation	Genetic and epigenetic changes	Associated genes	Function	References
Ionizing Radiation [γ-rays and x-rays]	Generation of free radicals by radiolysis of water and RNS		Attack other critical molecules of the cell	Azzam et al. (2012)
	Nitric oxide↑ Peroxynitrite (ONOO–)↑		Attack and damage thiols, DNA bases, proteins, etc.	Azzam et al. (2012)
			Cause DNA breaks, telomere dysfunction	
	ROS ↑ Superoxide anion (O$_2$–) ↑		Leads to: – Mutations in the mitochondrial DNA and alterations in the expression of proteins	Chen et al. (2003), Chaudhry and Omaruddin (2011), Choi et al. (2007)
			– Changes in the copy number	Malakhova and Bezlepkin (2005)
			– Genomic instability, DNA mutations	Antipova et al. (2011)
			– Autophagy	Lomonaco et al. (2009), Chiu et al. (2011)
	Hypomethylation↑ Hypermethylation↑ Global genome DNA methylation↑ Loss of methylation pattern ↓ in DNA expression levels of methyl-transferases (DNMTs) Methyl CpG binding proteins (MeCP2) ↓		Could result in genomic instability in the exposed animals	Kim et al. (2013), Raiche et al. (2004), Loree et al. (2006)

(continued)

Table 1 (continued)

Type of Radiation	Genetic and epigenetic changes	Associated genes	Function	References
	Methylation and Acetylation of lysine residues on histones—H3 and H4 and polycomb complex components H3K9acetyl ↓ H3K4me ↓ H3K9 ↑ H3K27 ↑ Histone phosphorylation at serine139 (γH2AX): γH2AX ↑			Jenuwein and Allis (2001)
				Nguyen et al. (2002), McGarvey et al. (2006)
				McGarvey et al. (2006)
			Main component for the measure of DSBs accumulation in the irradiated cells	Bonner et al. (2011), Pilch et al. (2003), Rogakou et al. (1998)
	Loss of functions of miRNAs ↓		Initiate and contribute to the malignant transformation of a normal cell	Nakamura et al. (2007), Calin et al. (2004), Saito et al. (2006)
	ATM phosphorylates SMC1 and activates it	BRCA1 NSB1 RAD9 RAD17 P53 CHK2	SMC1 plays an important role in controlling DNA replication forks and DNA repair after the damage.	Chistiakov et al. (2008)
	RAD51 shows DNA-dependent ATPase activity and has the ability to bind to the single and double stranded DNA. RAD51 unwinds the duplex DNA and forms helical nucleoprotein filaments at the site of DNA break RAD51 ↑	BRCA2	RAD51 is involved and leads to the activation of DSB and HR repair.	Chistiakov et al. (2008)
			RAD51 recombinase may play a role in increased risk of tumorigenesis	Kadouri et al. (2004), Levy-Lahad et al. (2001), Raderschall et al. (2002)

(continued)

Table 1 (continued)

Type of Radiation	Genetic and epigenetic changes	Associated genes	Function	References
	The kinases such as RAD-3 related, ATM, checkpoint kinase 2 phosphorylates BRCA1	RAD3 ATM CHK2	BRCA1 plays an important role in the promotion and regulation of HR Plays an important role in signal transduction induced by DNA damage	Chistiakov et al. (2008)
	BRCA1 and BRCA2 binds to RAD51 via BRC repeats	RAD51 recombinase	Mutations in BRCA1 and BRCA2 is strongly associated with the development of breast and ovarian cancer	Pellegrini et al. (2002), Honrado et al. (2005)
			The core HR mechanism is regulated by BRCA1 and BRCA2 via RAD51 recombinase	
	MRN complex binds to the broken ends of DNA and undergoes a series of conformational changes that leads to the activation of ATM.	RAD50, MRE11 and NBS1	MRN complex plays a major role in the detection of DNA damage and DNA damage response activation.	Paull and Lee (2005)
	DNA-PK is phosphorylated	XRCC4, Artemis, Ku70, Ku80, H2AX	DNA-PK becomes active and phosphorylates other proteins participating in DNA repair pathway	Collis et al. (2005)

(continued)

Table 1 (continued)

Type of Radiation	Genetic and epigenetic changes	Associated genes	Function	References
	XRCC4/DNA ligase IV have the capability to ligate one strand and thus the remaining SSB can be repaired as a single-stranded lesion.	The core components of NHEJ complex involves DNA-PK, DNA ligase IV and XRCC4	Radiosensitivity and genomic instability is promoted if there are genetic alterations in XRCC4 and DNA ligase IV	Chistiakov et al. (2008)
	DNA-PK phosphorylates Artemis	DNA-PK Artemis ATM	It activates the hairpin activity of Artemis which is essential for V(D)J recombination	Ma et al. (2002)
	Interaction of Ku and DNA polymerase μ	DNA polymeraseμ	Plays a major role in radiation-induced DNA repair in a manner dependent on the XRCC4/DNA ligase IV complex	Chistiakov et al. (2008)
Non-ionizing radiation (UV, radio, infrared, microwave)	UVR: Tumor suppressor gene p53 ↑ CPDs ↑ 6-4PP ↑ Pheomelanins cause oxidative damage by generation of free radicals in melanocytes		This mutation is beneficial for the cell growth and prevents the cell from undergoing apoptosis	Rebel et al. (2012)
			Cause mutations in DNA	Mallet et al. (2016)
			Causes DNA injury	Benedetto et al. (1982), Prota (2000), Sealy et al. (1982), Mitra et al. (2012)
	Direct (UVB) ROS ↑ CPDs ↑ 6-4PP ↑		Halts transcription and replication Produce DNA DSBs	Mallet et al. (2016), Batista et al. (2009), Limoli et al. (2002), Takahashi and Ohnishi (2005)

(continued)

Table 1 (continued)

Type of Radiation	Genetic and epigenetic changes	Associated genes	Function	References
	Indirect (UVA) ROS ↑ Eg: superoxide anions,		Causesoxidative DNA damage	Takeuchi et al. (1998) Meyskens et al. (2001)
	Hydrogen peroxide Oxidation of Nucleotide bases		Promotes mispairing among the bases causing mutations	Arvelo et al. (2016)
	Formation of oxidative products of purine and pyrimidine bases			Ravanat et al. (2001)

8.2 Sunscreen and Ultraviolet Radiation Tolerance

Sunscreen acting as the primary protector against the ultraviolet radiation plays a very important role. It has been observed that after using sunscreen the incidence rate of skin cancer and the aging process is reduced. By using the daily care and use of broad-spectrum sunscreens with the sun protection factor (SPF), the disastrous effect of frequent sub-erythematic exposure to human skin can be prevented. If sunscreen is applied to the skin, then repeated Ultraviolet radiation in small doses causes no harm to the texture of the skin (Chao et al. 2010).

8.3 Using Antioxidants

Use of antioxidants can facilitate preventing the generation of ROSand further inflammatory reactions caused by them. Topical utilization of vitamin E, carotenoids and ascorbic acid, which are naturally occurring cancer preventing antioxidants, play a major role in protection from the induced radiation (Pillai et al. 2005).

References

Agarwal S, Tafel AA, Kanaar R (2006) DNA double-strand break repair and chromosome translocations. DNA Repair 5(9–10):1075–1081. https://doi.org/10.1016/j.dnarep.2006.05.029
Alberto B, Chance B (1973) The mitochondrial generation of hydrogen peroxide: general properties and effect of hyperbaric oxygen. Biochem J 134(3):707–716. https://doi.org/10.1042/bj1340707

Antipova V, Malakhova V, Bezlepkin L, Vladimir B (2011) Detection of large deletions of mitochondrial DNA in tissues of mice exposed to X-rays. Biofizika 56:439–445

Armstrong BK, Kricker A (1993) How much melanoma is caused by sun exposure. Send Melanoma Res 6:395–401

Arvelo F, Sojo F, Cotte C (2016) Tumour progression and metastasis. Ecancermedicalscience 3:1–25. https://doi.org/10.3332/ecancer.2016.617

Azzam EI, Jay-Gerin JP, Pain D (2012) Ionizing radiation-induced metabolic oxidative stress and prolonged cell injury. Cancer Lett 327(1–2):48–60

Babior BM (1999) NADPH oxidase: an update. Blood 93:1464–1476

Bartel DP (2004) MicroRNAs: genomics, biogenesis, mechanism, and function. Cell 116(2):281–297

Batista LF, Kaina B, Meneghini R, Menck CF (2009) How DNA lesions are turned into powerful killing structures: insights from UV-Induced Apoptosis. Mutat Res—Rev Mutat Res 681(2–3):197–208. https://doi.org/10.1016/j.mrrev.2008.09.001

Benedetto JP, Ortonne JP, Voulot C, Khatchadourian C, Prota G, Thivolet J (1982) Role of thiol compounds in mammalian melanin pigmentation. II. glutathione and related enzymatic activities. J Invest Dermatol 79(6):422–424. https://doi.org/10.1111/1523-1747.ep12530631

Bernstein E, Allis CD (2005) RNA meets chromatin. Genes Dev 212:1635–1655. https://doi.org/10.1101/gad.1324305.GENES

Berwick M, Wiggins C (2006) The current epidemiology of cutaneous malignant melanoma. Send to Front Biosci 11:1244–1254

Bonner WM, Redon CE, Dickey JS, Nakamura AJ, Sedelnikova OA, Solier S, Pommier Y (2011) Nat Rev Cancer 8(12):957–967. https://doi.org/10.1038/nrc2523

Boveris A, Cadenas E, Stoppani AO (1976) Role of ubiquinone in the mitochondrial generation of hydrogen peroxide. Biochem J 156(2):435–444. https://doi.org/10.1042/bj1560435

Cadenas E, Davies KJ (2000) Mitochondrial free radical generation, oxidative stress, and aging. Free Radic Biol Med 29:222–230

Calin GA, Sevignani C, Dumitru CD, Hyslop T, Noch E, Yendamuri S, Shimizu M et al (2004) Human MicroRNA genes are frequently located at fragile sites and genomic regions involved in cancers. Proc Natl Acad Sci 101(9):2999–3004. https://doi.org/10.1073/pnas.0307323101

Celeste A, Fernandez-Capetillo O, Kruhlak MJ, Pilch DR, Staudt DW, Lee A, Bonner RF, Bonner WB, Nussenzweig A (2003) Histone H2AX phosphorylation is dispensable for the initial recognition of DNA breaks. Nat Cell Biol 5(7):675–679. https://doi.org/10.1038/ncb1004

Chan DW, Chen BPC (2002) Autophosphorylation of the DNA-dependent protein kinase catalytic subunit is required for rejoining of DNA double-strand breaks. Genes Dev 16:2333–2338. https://doi.org/10.1101/gad.1015202.We

Chao Y, Xue-Min W, Yi-Mei T, Li-Jie Y, Yin-Fen L, Pei-Lan W (2010) Effects of sunscreen on human skin's ultraviolet radiation tolerance. J Cosmet Dermatol 9(4):297–301. https://doi.org/10.1111/j.1473-2165.2010.00525.x

Chaudhry MA, Omaruddin RA (2011) Mitochondrial gene expression in directly irradiated and nonirradiated bystander cells. Cancer Biother Radiopharm 26:657–663

Chen Q, Chai YC, Mazumder S, Jiang C, Macklis R, Chisolm G, Almasan A (2003) The late increase in intracellular free radical oxygen species during apoptosis is associated with cytochrome c release, caspase activation, and mitochondrial dysfunction. Cell Death Differ 10(3):323–334. https://doi.org/10.1038/sj.cdd.4401148

Chistiakov DA, Voronova NV, Chistiakov PA (2008) Genetic variations in DNA repair genes, radiosensitivity to cancer and susceptibility to acute tissue reactions in radiotherapy-treated cancer patients. Acta Oncol 47(5):809–824. https://doi.org/10.1080/02841860801885969

Chiu HW, Lin W, Ho SY, Wang YJ (2011) Synergistic effects of arsenic trioxide and radiation in osteosarcoma cells through the induction of both autophagy and apoptosis. Radiat Res 175:547–560

Choi KM, Kang CM, Cho ES, Kang SM, Lee SB, Um HD (2007) Ionizing radiation-induced micronucleus formation is mediated by reactive oxygen species that are produced in a manner dependent on mitochondria, Nox1, and JNK. Oncol Rep 17(5):1183–1188

Collis SJ, DeWeese TL, Jeggo PA, Parker AR (2005) The life and death of DNA-PK. Oncogene 24(6):949–961. https://doi.org/10.1038/sj.onc.1208332

D'Arcangelo M, Drew Y, Plummer R (2016) The role of PARP in DNA repair and its therapeutic exploitation. DNA repair in cancer therapy: molecular targets and clinical applications: second edition 105(8):115–134. https://doi.org/10.1016/B978-0-12-803582-5.00004-8

Droge W (2002) Free radicals in the physiological control of cell function. Am Physiol Soc 82:47–95

Du C, Gao Z, Venkatesha VA, Kalen AL, Chaudhuri L, Spitz DR, Cullen JJ, Oberley LW, Goswami PC (2009) Mitochondrial ROS and radiation induced transformation in mouse embryonic fibroblasts. Cancer Biol Ther 8(20):1962–1971. https://doi.org/10.4161/cbt.8.20.9648

Fan R, Kumaravel TS, Jalali F, Marrano P, Squire JA, Bristow RG (2004) Defective DNA strand break repair after DNA damage in prostate cancer cells: implications for genetic instability and prostate cancer progression. Can Res 64(23):8526–8533. https://doi.org/10.1158/0008-5472.CAN-04-1601

Hasselbach L, Haase S, Fischer D, Kolberg HC, Stürzbecher HW (2005) Characterisation of the promoter region of the human DNA-repair gene Rad51. Eur J Gynaecol Oncol 26(6):589–598

Heale JT, Alexander RB, Schmiesing JA, Kim JS, Kong X, Zhou S, Hudson DF, Earnshaw WC, Yokomori K (2006) Condensin I interacts with the PARP-1-XRCC1 complex and functions in DNA single-strand break repair. Mol Cell 21(6):837–848. https://doi.org/10.1016/j.molcel.2006.01.036

Hein AL, Ouellette MM, Yan Y (2014) Radiation-induced signaling pathways that promote cancer cell survival (review). Int J Oncol 45(5):1813–1819. https://doi.org/10.3892/ijo.2014.2614

Honrado E, Benítez J, Palacios J (2005) The molecular pathology of hereditary breast cancer: genetic testing and therapeutic implications. Mod Pathol 18(10):1305–1320. https://doi.org/10.1038/modpathol.3800453

Ilnytskyy Y, Zemp FJ, Koturbash I, Kovalchuk O (2008) Altered MicroRNA expression patterns in irradiated hematopoietic tissues suggest a sex-specific protective mechanism. Biochem Biophys Res Commun 377:41–45

Ito S, Wakamatsu K, Ozeki H (2000) Chemical analysis of melanins and its application to the study of the regulation of melanogenesis. Pigm Cell Res/Sponsored Eur Soc Pigm Cell Res Int Pigm Cell Soc 13(Suppl 8):103–109. https://doi.org/10.1034/j.1600-0749.13.s8.19.x

Jawad M, Seedhouse CH, Russell N, Plumb M (2012) Genes increase the risk of therapy-related acute myeloid leukemia brief report polymorphisms in human homeobox HLX1 and DNA repair RAD51 genes increase the risk of therapy-related acute myeloid leukemia. Blood 108(12):3916–3918. https://doi.org/10.1182/blood-2006-05-022921

Jenuwein T, Allis CD (2001) Allis: translating the histone code. Science 293(August):1074–1080

Kadouri L, Kote-Jarai Z, Hubert A, Durocher F, Abeliovich D, Glaser B, Hamburger T, Eeles RA, Peretz T (2004) A single-nucleotide polymorphism in the RAD51 gene modifies breast cancer risk in BRCA2 carriers, but not in BRCA1 carriers or noncarriers. Br J Cancer 90(10):2002–2005. https://doi.org/10.1038/sj.bjc.6601837

Kawada A (2000) Risk and preventive factors for skin phototype. J Dermatol Sci 23:527–529

Kim JG, Park MT, Heo K, Yang KM, Yi J (2013) Epigenetics meets radiation biology as a new approach in cancer treatment. Int J Mol Sci 14(7):15059–15073. https://doi.org/10.3390/ijms140715059

Kryston TB, Georgiev AB, Pissis P, Georgakilas AG (2011) Role of oxidative stress and DNA damage in human carcinogenesis. Mutat Res 711:193–201

Leach JK, Glenn VT, Lin P, Schmidt-ullrich R, Mikkelsen RB (2001) Ionizing radiation-induced, mitochondria-dependent generation of reactive oxygen/nitrogen. Can Res 1:3894–3901

Levy-Lahad E, Lahad A, Eisenberg S, Dagan E, Paperna T, Kasinetz L, Catane R et al (2001) A single nucleotide polymorphism in the RAD51 gene modifies cancer risk in BRCA2 but not BRCA1 carriers. Proc Nat Acad Sci 98(6):3232–3236. https://doi.org/10.1073/pnas.051624098

Limoli CL, Giedzinski E, Bonner WM, Cleaver JE (2002) UV-induced replication arrest in the xeroderma pigmentosum variant leads to DNA double-strand breaks, γH2AX formation, and Mre11 relocalization. Proc Natl Acad Sci USA 99(1):233–238. https://doi.org/10.1073/pnas.231611798

Liu C, Vyas A, Kassab MA, Singh AK, Yu X (2017) The role of poly ADP-ribosylation in the first wave of DNA damage response. Nucleic Acids Research 45(14):8129–8141. https://doi.org/10.1093/nar/gkx565

Lomonaco SL, Finniss S, Xiang C, DeCarvalho A, Umansky F, Kalkanis SN, Mikkelsen T, Brodie C (2009) The induction of autophagy by? radiation contributes to the radioresistance of glioma stem cells. Int J Cancer 125(3):717–722. https://doi.org/10.1002/ijc.24402

Loree J, Koturbash I, Kutanzi K, Baker M, Pogribny I, Kovalchuk O (2006) Radiation-induced molecular changes in rat mammary tissue: possible implications for radiation-induced carcinogenesis. Int J Radiat Biol 82:805–815

Los M, Schenk H, Hexel K, Baeuerle PA, Dröge W, Schulze-Osthoff K (1995) IL-2 Gene expression and NF-kappa B activation through CD28 requires reactive oxygen production by 5-lipoxygenase. The EMBO J 14(15):3731–3740

Lu H, Edwards C, Gaskell S, Pearse A, Marks R (1996) Melanin content and distribution in the surface corneocyte with skin phototypes. Send to Br J Dermatol 2:263–267

Ma Y, Pannicke U, Schwarz K, Lieber MR (2002) Hairpin opening and overhang processing by an artemis/DNA-dependent protein kinase complex in nonhomologous end joining and V(D) J recombination. Cell 108(6):781–794. https://doi.org/10.1016/S0092-8674(02)00671-2

Malakhova L, Bezlepkin VG (2005) The increase in mitochondrial DNA copy number in the tissues of Γ-irradiated mice. Cell Mol Biol Lett 10:721–32. http://www.cmbl.org.pl

Mallet JD, Dorr MM, Desgarnier MCD, Bastien N, Gendron SP, Rochette PJ (2016) Faster DNA repair of ultraviolet-induced cyclobutane pyrimidine dimers and lower sensitivity to apoptosis in human corneal epithelial cells than in epidermal keratinocytes. PLoS ONE 11(9):1–22. https://doi.org/10.1371/journal.pone.0162212

McGarvey KM, Fahrner JA, Greene E, Martens J, Jenuwein T, Baylin SB (2006) Silenced tumor suppressor genes reactivated by DNA demethylation do not return to a fully euchromatic chromatin state. Can Res 66(7):3541–3549. https://doi.org/10.1158/0008-5472.CAN-05-2481

Meyskens FL, Farmer P, Fruehauf JP (2001) Redox regulation in human melanocytes and melanoma. Pigm Cell Res 14(3):148–154. https://doi.org/10.1034/j.1600-0749.2001.140303.x

Mitra D, Luo X, Morgan A, Wang J, Hoang MP, Lo J, Guerrero CR et al (2012) A UV-independent pathway to melanoma carcinogenesis in the redhair-fairskin background. Nature 491(7424):449–453. https://doi.org/10.1038/nature11624.A

Morrison C (2000) The controlling role of ATM in homologous recombinational repair of DNA damage. EMBO J 19(3):463–471. https://doi.org/10.1093/emboj/19.3.463

Nakamura T, Canaani E, Croce CM (2007) Oncogenic all1 fusion proteins target drosha-mediated microRNA processing. Proc Natl Acad Sci 104(26):10980–10985. https://doi.org/10.1073/pnas.0704559104

Narayanan DL, Saladi RN, Fox JL (2010) Ultraviolet radiation and skin cancer. Int J Dermatol 49(9):978–986. https://doi.org/10.1111/j.1365-4632.2010.04474.x

Nguyen CT, Weisenberger DJ, Velicescu M, Gonzales FA, Lin JCY, Liang G, Jones PA (2002) Histone H3-lysine 9 methylation is associated with aberrant gene silencing in cancer cells and is rapidly reversed by 5-aza-2'-deoxycytidine. Cancer Res 323:6456–6461

O'Neill P, Wardman P (2009) Radiation chemistry comes before radiation biology. Int J Radiat Biol 1:9–25

Orazio JD, Jarrett S, Amaro-ortiz A, Scott T (2013) UV radiation and the skin. Int J Mol Sci 14:12222–12248. https://doi.org/10.3390/ijms140612222

Pain D, Murakami H, Blobel G (1990) Identification of a receptor for protein import into mitochondria. Nature 347:444–449

Paull TT, Lee JH (2005) The Mre11/Rad50/Nbs1 complex and its role as a DNA double-strand break sensor for ATM. Cell Cycle 4(6):737–740. https://doi.org/10.4161/cc.4.6.1715

Pellegrini L, Yu DS, Lo T, Anand S, Lee MY, Blundell TL, Venkitaraman AR (2002) Insights into DNA recombination from the structure of a RAD51-BRCA2 complex. Nature 420(6913):287–293. https://doi.org/10.1038/nature01230

Peter AJ, Baylin SB (2010) The epigenomics of cancer. Omics Perspect Cancer Res 128(4):51–67. https://doi.org/10.1007/978-90-481-2675-0_4

Pilch DR, Sedelnikova OA, Redon C, Celeste A, Nussenzweig A, Bonner WM (2003) Characteristics of γH2AX foci at DNA double-strand breaks sites. Biochem Cell Biol 81:123–129

Pillai S, Oresajo C, Hayward J (2005) Ultraviolet radiation and skin aging : roles of reactive oxygen species, inflammation and protease activation, and strategies for prevention of inflammation-induced matrix degradation—a review. Int J Cosmet Sci 27:17–34

Pleasance ED, Cheetham RK, Stephens PJ, McBride DJ, Humphray SJ, Greenman CD, Varela I et al (2010) A comprehensive catalogue of somatic mutations from a human cancer genome. Nature 463(7278):191–196. https://doi.org/10.1038/nature08658

Pogribny I, Raiche J, Slovack M, Kovalchuk O (2004) Dose-dependence, sex-and tissue-specificity, and persistence of radiation-induced genomic DNA methylation changes. Biochem Biophys Res Commun 320:1253–1261

Poulton R, Caspi A, Milne BJ, Thomson WM, Taylor A, Sears MR, Moffitt TE (2013) Association between children's experience of socioeconomic disadvantage and adult health: a life-course study. Lancet 360(9346):1640–1645. https://doi.org/10.1016/S0140-6736(02)11602-3. Association

Prota G (2000) Melanins, melanogenesis and melanocytes: looking at their functional significance from the chemist's viewpoint. Pigm Cell Res 13(4):283–293. https://doi.org/10.1034/j.1600-0749.2000.130412.x

Raderschall E, Stout K, Freier S, Suckow V, Schweiger S, Haaf T (2002) Elevated levels of Rad51 recombination protein in tumor cells. Can Res 62(1):219–225

Raiche J, Rodriguez-Juarez R, Pogribny I, Kovalchuk O (2004) Sex- and tissue-specific expression of maintenance and de novo DNA methyltransferases upon low dose X-Irradiation in mice. Biochem Biophys Res Commun 325:39–47

Ramasamy K, Shanmugam M, Balupillai A, Govindhasamy K, Gunaseelan S, Muthusamy G, Robert BM, Nagarajan RP (2017) Ultraviolet radiation-induced carcinogenesis: mechanisms and experimental models. J Radiat Cancer Res 8(1):4. https://doi.org/10.4103/0973-0168.199301

Ravanat JL, Douki T, Cadet J (2001) Direct and indirect effects of uv radiation on DNA and its components. J Photochem Photobiol, B 63(1–3):88–102. https://doi.org/10.1016/S1011-1344(01)00206-8

Ravnbak MH (2010) Objective determination of fitzpatrick skin type. Dan Med Bull 57:B4153

Rebel HG, Bodmann CA, van de Glind GC, de Gruijl FR (2012) UV-induced ablation of the epidermal basal layer including P53-mutant clones resets UV carcinogenesis showing squamous cell carcinomas to originate from interfollicular epidermis. Carcinogenesis 33(3):714–720. https://doi.org/10.1093/carcin/bgs004

Rogakou EP, Pilch DR, Orr AH, Ivanova VS, Bonner WM (1998) Double-stranded breaks induce histone H2AX phosphorylation on serine 139. J Biol Chem 273(10):5858–5868. https://doi.org/10.1074/jbc.273.10.5858

Rogers HW, Weinstock MA, Harris AR, Hinckley MR, Feldman SR, Fleischer AB, Coldiron BM (2010) Incidence estimate of nonmelanoma skin cancer in the United States. Arch Dermatol 146(1538):283–287

Sahin E, Colla S, Liesa M, Moslehi J, Müller FL, Guo M, Cooper M et al (2011) Telomere dysfunction induces metabolic and mitochondrial compromise. Nature 470(7334):359–365. https://doi.org/10.1038/nature09787

Saito Y, Liang G, Egger G, Friedman JM, Chuang JC, Coetzee GA, Jones PA (2006) Specific activation of microRNA-127 with downregulation of the proto-oncogene BCL6 by chromatin-modifying drugs in human cancer cells. Cancer Cell 9(6):435–443. https://doi.org/10.1016/j.ccr.2006.04.020

Scherer D, Kumar R (2010) Genetics of pigmentation in skin cancer—a review. Send to Mutat Res 2:141–153

Sealy RC, Hyde JS, Felix CC, Menon IA, Prota G, Swartz HM, Persad S, Haberman HF (1982) Novel free radicals in synthetic and natural pheomelanins: distinction between dopa melanins and cysteinyldopa melanins by ESR spectroscopy. Proc Natl Acad Sci USA 79(9):2885–2889. https://doi.org/10.1073/pnas.79.9.2885

Seedhouse C, Faulkner R, Ashraf N, Das-Gupta E, Russell N (2004) Polymorphisms in genes involved in homologous recombination repair interact to increase the risk of developing acute myeloid leukemia. Clin Cancer Res 10(8):2675–2680. https://doi.org/10.1158/1078-0432.CCR-03-0372

Singh PK, Campbell MJ (2013) The interactions of microRNA and epigenetic modifications in prostate cancer. Cancers (Basel) 5(3):998–1019

Slominski A (2004) Melanin pigmentation in mammalian skin and its hormonal regulation. Physiol Rev 84(4):1155–1228. https://doi.org/10.1152/physrev.00044.2003

Soldatenkov VA, Smulson M (2000) Poly (ADP-ribose) polymerase in DNA damage-response pathway: implications for radiation oncology. Int J Cancer 90(2):59–67. https://doi.org/10.1002/(SICI)1097-0215(20000420)90:2%3c59:AID-IJC1%3e3.0.CO;2-4

Soubeyrand S, Pope L, Chasseval RD, Gosselin D, Dong F, de Villartay JP, Haché RJG (2006) Artemis phosphorylated by DNA-dependent protein kinase associates preferentially with discrete regions of chromatin. Journal of Molecular Biology 358(5):1200–1211. https://doi.org/10.1016/j.jmb.2006.02.061

Spitz DR, Azzam EI, Li JJ, Gius D (2004) Metabolic oxidation/reduction reactions and cellular responses to ionizing radiation: a unifying concept in stress response biology. Cancer Metastasis Rev 23:311–322

Takahashi A, Ohnishi T (2005) Does γH2AX foci formation depend on the presence of DNA double strand breaks? Send to Cancer Lett 2:171–179

Takeuchi T, Uitto J, Bernstein EF (1998) A novel in vivo model for evaluating agents that protect against ultraviolet a-induced photoaging. Journal of Investigative Dermatology 110(4):343–347. https://doi.org/10.1046/j.1523-1747.1998.00124.x

Thacker J (2005) The RAD51 gene family, genetic instability and cancer. Cancer Lett 219(2):125–135. https://doi.org/10.1016/j.canlet.2004.08.018

Valerie K, Yacoub A, Hagan MP, Curiel DT, Fisher PB, Grant S, Dent P (2007) Radiation-induced cell signaling: inside-out and outside-in. Mol Cancer Ther 6(3):789–801. https://doi.org/10.1158/1535-7163.MCT-06-0596

Vincensi MR, d'Ischia M, Napolitano A, Procaccini EM, Riccio G, Monfrecola G, Santoianni P, Prota G (1998) Phaeomelanin versus eumelanin as a chemical indicator of ultraviolet sensitivity in fair-skinned subjects at high risk for melanoma: a pilot study. Send to Melanoma Res 1:53–58

Wei H, Yu X (2016) Functions of PARylation in DNA damage repair pathways. Genom Proteomics Bioinform 14(3):131–139. https://doi.org/10.1016/j.gpb.2016.05.001

Wondrak GT, Jacobson MK, Jacobson EL (2006) Endogenous UVA-photosensitizers: mediators of skin photodamage and novel targets for skin photoprotection. Photochem Photobiol Sci 5(2):215–237. https://doi.org/10.1039/B504573H

Yarosh D, Klein J, Kibitel J, Alas L, O'Connor A, Cummings B, Grob D et al (1996) Enzyme therapy of xeroderma pigmentosum: safety and efficacy testing of T4N5 liposome lotion containing a prokaryotic DNA repair enzyme. Photodermatol Photoimmunol Photomed 12(3):122–130. https://doi.org/10.1111/j.1600-0781.1996.tb00188.x

Zhang X, Succi J, Feng Z, Story M, Legerski RJ, Prithivirajsingh S (2004) Artemis is a phosphorylation target of ATM and ATR and is involved in the G2/M DNA damage checkpoint response artemis is a phosphorylation target of ATM and ATR and is involved in the G2/M DNA damage checkpoint response. Mol Cell Biol 24(20):9207–9220. https://doi.org/10.1128/MCB.24.20.9207

Mechanistic Effect of Heavy Metals in Neurological Disorder and Brain Cancer

Sandeep Kumar Agnihotri and Kavindra Kumar Kesari

Abstract Industrialization era is considered as a part of important human development. Industrialization increases the extensive use of different metals from earth crust because of their materials demand. Extensive use of these materials in daily life and their improper disposal are the reasons for environmental pollution. Toxic metals are highly causative in an open environment and because of this human gets exposures frequently. These toxic metal like cadmium (Cd), lead (Pb), Arsenic (As), Mercury (Hg), Thallium (Th) cross the blood brain barrier to enter into the brain and leads to development of neurodegenerative diseases. Heavy metals play an important role by inducing the reactive oxygen species, mitochondrial dysfunction, calcium ion efflux, an activation of immunogenic response, and suppression of anti-oxidants like catalase, superoxide dismutase (SOD), glutathione. Moreover, the brain-derived neurotrophic factor (BDNF) causes the depletion in cognitive dysfunctions and impairs the memory functions with several other neurological diseases like Alzheimer's and Parkinson's diseases. Here we have tried to illustrate the metals evoked mechanism, which impaires the function of neurons and generate the neurotoxicity and neurodegenerative diseases.

Keywords Cancer · Heavy metals · Neurotoxicty · Neuronal diseases

1 Introduction

The environment has started polluting after the industrialization and this led to contamination of water, air and whole atmosphere with unnatural gases. Pollution disturbs the ecosystem with all forms of industrial waste in the modern age of human civilization and development, pollution level is at its peak and responsible for severe human illness and diseases. Toxic environment affects more than 200 million people

S. K. Agnihotri (✉)
School of Life Science and Technology, Harbin Institute of Technology, Harbin, China
e-mail: sandeepagnihotri33@yahoo.com

K. K. Kesari
Department of Applied Physics, Aalto University, Espoo, Finland

© Springer Nature Switzerland AG 2019　　　　　　　　　　　　　　　　　　　25
K. K. Kesari (ed.), *Networking of Mutagens in Environmental Toxicology*, Environmental Science,
https://doi.org/10.1007/978-3-319-96511-6_2

worldwide and its worst effects are on newborn babies. Babies' health and development are impaired the most because pollution impact when immunity is weak and a recent UNICEF report shows that pollution causes the defect in development of the brain of babies along with causing cardiovascular and respiratory diseases.

Metals are very important for the living system and their homeostasis. However, if homeostasis collapsed, metal binds to the different site of protein (Nelson 1999) may impaired and lead to cause different kind of diseases (Halliwell and Gutteridge 2007). Metals also affect the gene regulation (Arini et al. 2015) and signaling pathways, which may be responsible for the cell growth and differentiation (Christie et al. 2013). However, deregulation of cell growth and differentiation leads to the cancer and apoptosis growth (Tykwinska et al. 2013).

Heavy metal is used in many occupations like, welding, smelting, pipe industries, painting, household utensils industries and many innumerable industries. Heavy metal has become a big problem with intractable outcome. This is affecting every individual in a direct or indirect way and leads to the varied diseases related to cardiovascular, respiratory, reproduction and cancers (Carpenter and Jiang 2013). Heavy metals are associated with neurodegenerative diseases such as multiple sclerosis, amyotrophic lateral sclerosis, Parkinson's disease, Wilson disease and Alzheimer's disease (Giacoppo et al. 2014; Dusek et al. 2015).

Natural environment is a habitat for all biological living organisms. Most of the biological organisms have been evolving with the ages and show enormous development in respect to their previous generations. Humans belong to this category and the enormous level of development in all the areas makes life stress-free. Progress of human society increased rapidily with an industrialization in the 18th century. In the expansion of industrialization, we have builded roads for transportation, motor vehicles to transport materials, and for this, we have extracted the materials from the Earth crust and instigated the environmental damage. As the industrialization progresses, the spreading of wastes materials started polluting the atmosphere. These waste materials comprises an enormous amount of heavy metals, which gradually enters into our body system, accumulate and impaires the body homeostasis. Homeostasis is necessary to maintain the inner physiological activity and body organ functions. However, any kind of disruption due to heavy metals may start damaging the function of organs. Heavy metals commenced to disrupt the functions of all the organs including the brain (Agnihotri et al. 2015). Brain is an organ of our central nervous system, which is a most complex and it comprises of networks of billions of neurons, responsible for homeostatic activity of body, behavior, mood, muscular activity, memory etc. Brain functions affected by heavy metals may induce toxicity to interrupt the regular functions of central nervous system (Zhang et al. 2016). Heavy metals role has been understood in major neurobiological diseases such as Parkinson's disease, Alzheimer's disease, autism, multiple sclerosis (MS), amyotrophic lateral sclerosis/motor neuron disease (ALS/MND), dementia and cognitive disorder (Notarachille et al. 2014; Kumudini et al. 2014; Carocci et al. 2014; Caffo et al. 2014) (Fig. 1).

Heavy metal toxicants exists in dissimilar form in the environment and introduced in a human body by water, air and food. It is difficult to degrade the heavy metal by

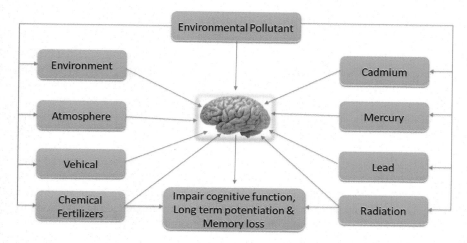

Fig. 1 Brain is very important part of our body, and is extremely sensitive too. Environmental pollutant and other heavy metals impact on brain and impairs the cognitive functions, causing long term potentiation and memory loss, and also impairing an important part of decision making

existing proteins and enzymes in the body. Therefore, proteins (metallothionein) bind with heavy metal and excrete from the body and reduce the toxicity of heavy metal (Liu et al. 2014). However, heavy metal dissolves in blood and circulate through different organs of body that persuades signaling mechanism to generate toxicity. Brain has specific respect to other organs of the body because; it has blood brain-barriers (BBB), which separates the circulating blood from brain and extracellular fluid of the central nervous system (CNS). Blood brain barrier allows the passage of glucose and amino acid by some selective transport mechanism, however, water, gases and lipid soluble molecules may transported with passive diffusion. Blood brain barrier is a vital part of brain to prevent neurodegeneration from external toxicants (Zheng et al. 2003) and is used as a defense mechanism for the brain (Nathanson and Mischel 2011). Heavy metal impairs the blood brain barrier and induces the oxidative stress, which instigates the cell death and initiates the neurodegeneration through different signaling pathway (Caserta et al. 2013).

Toxicity of heavy metal pollutant is enormously hazardous for developing central nervous system. In offspring's, small amount of intoxication affect their ability to learn and memorize their cognitive ability. At the developmental stage, low level of heavy metal impedes the development of the brain and generates neurodegenerative diseases likes' autism but the mechanism is completely obscure (Mohamed Fel et al. 2015; Yassa 2014). Here, we have discussed about the impact of heavy metals on neurodegenerative diseases and related mechanism.

2 Heavy Metal and Blood Brain Barrier

Etiology of neurobiological disorder is not completely attributed to acquired behaviors, proposing that heavy metals and other environmental factors possibly contribute too. Blood brain barrier (BBB) has been involved in metal transport and neural defense mechanism. BBB transport energy (glucose and amino acid) and important ions to maintain the homeostasis of neurons and glial cells and excrete the waste materials from the brain. Metal ion is exchanged extremely slowly between plasma and brain compared to other tissues (Serlin et al. 2015). Metals deposits with an accumulation of proteins have instigated the inflammatory process near the endothelial wall of BBB (Joana et al. 2016; McCarthy et al. 2018) within brain parenchyma, which leads to neuronal damage and loss (Cherry et al. 2014; Greter and Merad 2013). This progression of the brain degeneration for an extended period leads to disability and demise of neural cells. Alzheimer disease and Parkinson diseases are old age diseases that reflect the inflammation caused by the metals and accumulated proteins (Zeineh et al. 2015; Heppner et al. 2015). These metals are found in many forms around us since birth and accumulate in brain slowly as aluminum (Al), copper (Cu), iron (Fe), manganese (Mn) lead (Pb), mercury (Hg) and zinc (Zn) as an essential and non-essential manner. It has been noticed that impede mechanism of homeostasis triggered by metals indicate the sporadic form of Alzheimer disease (Li et al. 2015; Grubman et al. 2014). Al, Cu, Fe and Zn are the metals identified for sporadic form of Alzheimer disease (Grubman et al. 2014). Fe and Cu deficit have been analyzed in the substantia nigra of Parkinson's diseases (Loef and Walach 2015). Epidemiological studies suggests that Al, Cu, Fe, Pb, Mn, Hg and Zn are risk factors for Parkinson's disease (Salvador et al. 2010; Zucca et al. 2017; Greenough et al. 2013; Meyer et al. 2015; Doorn and Kruer 2013). Amyotrophic Lateral Sclerosis (ALS) is a motor neuron disease in which death of neurons controlling the voluntary muscle in around 20% of cases is caused by mutation of Cu, Zn-superoxide dismutase (SOD1) (Desai and Kaler 2008; Roos et al. 2006). ALS patient have minimum amount of Ca and Mg and elevated amount of Al, Cu, Pb, Mn and Hg (Fondell et al. 2013; Robison et al. 2015).

Heavy metals may enters in our body system through polluted air, contaminated water and food products and absorbed by intestine, lungs, and also through skin. Further, they mixed in blood and circulated to central nervous system (CNS) by crossing the BBB or choroid plexus (CP) to cerebrospinal fluid (CSF) or by diffusion to CNS (Yokel 2006). Although few other metals can be absorbed by a sensory nerve in the nasal cavity and enters into the brain (Oliver and Fazakerley 1998; Dorman et al. 2002). Mn and Ni also enters through nasal cavity (Rao et al. 2003). Moreover, glucose is the primary energy substrate of the brain, and its metabolism accounts for nearly all of the oxygen consumption in the brain (Mergenthaler et al. 2013). Brain glucose demands greatly and exceeds the rate of glucose diffusion across the BBB, where, Glut-1 mediates the brain glucose uptake to meet the brain's needs (Bélanger et al. 2011; Camandola and Mattson 2017). Glucose metabolized by the brain is of two or three times higher (Hyder et al. 2006) than any other organs of the body. If

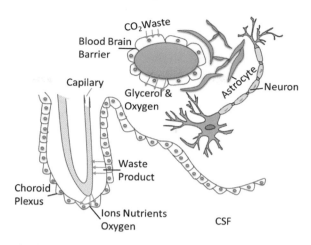

Fig. 2 Tight junction between epithelial cells make the blood brain barrier between capillary and cerebrospinal fluid, which passes the ions, nutrients, oxygen from blood capillary to cerebrospinal fluid (CSF) and remove the waste from CSF to capillary, in ventricle of the choroid plexus. BBB exchange materials and gas by diffusion or by specific transporter from blood to brain. Astrocytes remove the toxicants through BBB and take glucose and oxygen from blood and supply to neurons

BBB is damaged and compromised with energy delivery in the brain, that it may leads to seizures, mental retardation, compromised brain development and low CSF glucose concentrations in children (Yang et al. 2013) (Fig. 2).

Metals also transported by some transporter proteins as divalent metal transporter (DMT1, DCT1, Nramp2), which transports only divalent metals (Ca, Fe, Mn, Cd, Cu, Ni, and Pb) (Harris 1983). However transporter proteins as transferrin (transferrin mediated endocytosis) binds to metal and transported to CSF passing through BBB (Yokel 2006), diffusion, fluid phase endocytosis, receptor mediated endocytosis.

Aluminium (Al) reaches into the brain after crossing the BBB through transferrin-mediated endocytosis (TfR-ME) within 4 hours, when it passes into the body as aluminium citrate (Davson et al. 1987). Copper is transported by ATP receptors (ATP7A, ATP7B and CTR1) at BBB. Fe is also transported by (1) transferrin-receptor mediated endocytosis, (2) non-transferrin dependent mechanism, (3) DMT1 at BBB to the brain. Transportation of lead (Pb) through BBB in the brain is done by passive diffusion as $PbOH^+$ and cation channel (ATP dependent Ca pump). Mercury is transported only as methyl mercury (MeHg), not in any other form (Aschner 2007), but some study suggests that it makes complex with L-cysteine and transported through L-system (Aschner and Clarkson 1989). Mn is passed through BBB by diffusion, and transferrin mediated endocytosis to reach the brain where, Zn transported by Zip1 and Zip 6 at the BBB. Cd enters via the olfactory bulb and goes through central neurons, and damages the BBB permeability (Chowanadisai et al. 2005). BBB crossed metals reached to CSF and effluxes by astrocytes (protect the brain by taking out the

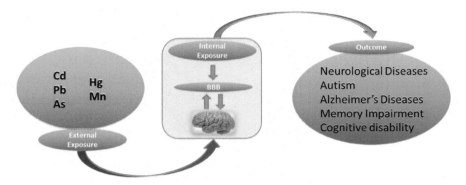

Fig. 3 Representation of different metals, present in the environment crossed the blood brain barrier (Ankley et al. 2010) and evoked neurodegenerative diseases with memory impairment and cognitive deficit

leftover and surplus material) to prevent any oxidative damage or reaction, which instigates numerous signaling processes to develop neurological diseases (Fig. 3).

3 Heavy Metal and Neurological Disease

Metals are divided in two categories, essential and non-essential metals, according to need of the living organism. Essential metals comprise the property to modulate the enzymatic and cellular activity, which controls the physiological actions such as protein reformation, signaling pathways, electron transport, redox reactions, metabolism of protein (carbohydrate and Fat), transportation of molecules, immune reactions and neurotransmitter synthesis and transportation (Wu et al. 2018; Lopez et al. 2002; Giralt et al. 2000; Murakami and Hirano 2008). These metals are essential for cellular lifespan, however, deficiencies of these metals linked to several neurodegenerative diseases. Moreover, the elevated levels may also induces the intracellular process, and causes the apoptosis, autophagy, mitophagy, signal transduction, protein misfolding, oxidative stress, mitochondrial dysfunction (Wegst et al. 2015; Szabo et al. 2016; Peres et al. 2016). This might be vulnerable to the normal brain functioning and cause many neurodegenerative diseases. Metals also affects the memory dysfunction, cognitive ability, muscular dysfunction, neurons apathy, amyotrophic lateral sclerosis, huntington's disease, menkes disease, Wilson's Disease, Friedreich's ataxia, gulf war syndrome, manganese multiple sclerosis, autism, insomnia, anorexia, anxiety, depression, Alzheimer's disease and Parkinson's disease (Sparks and Schreurs 2003; Mohandas and Colvin 2004; Koppenal et al. 2004; Andrade et al. 2017; Caito and Aschner 2015).

Aluminium (Al) reached into body or different organs through drinking water and food. Epidemiological survey identified that higher concentration of Al in drinking water may develops dementia, like Alzheimer's disease (Yokel and Florence 2006;

White et al. 1992; Killin et al. 2016). Exposure to miner concentration (inhale fine Al & Al_2O_3) may cause cognitive impairment, slow psychomotor response, and memory loss (Hosovski et al. 1990). Moreover, several other diseases such as, fatigue, working memory and learning behaviors were found in shipyard Al welders (Riihimaki et al. 2000). However, the role of Al is controversial for AD, because several other studies reveals that AD patients had the same amount of Al as control group in the urine and plasma (Graves et al. 1998). Study also suggests that changes occurs in AD brain were the same as found in AD (Perl 2001), where, Al increases the plaque deposition, Aβ protein aggregation and polymerization, and Aβ production in the brain (Clauberg and Joshi 1993; Mantyh et al. 1993; Kawahara et al. 1994). Aluminium hydroxide impede the neurological functions and persuade autism, increase anxiety, depression, long-term memory loss and neuronal death in spinal and motor cortex (Shaw and Tomljenovic 2013). Brain impairment depends on the exposure of Al concentrations and period causes physiological changes, which appear later in life. Manganese (Mn) mostly used in fuel additive, to reduce the combustion of the engine and released in air as manganese sulphate and phosphate (Lynam et al. 1999). Therefore the environment polluted air (manganese sulphate and phosphate) ingested by humans lead to the brain exposures. Mn may interrupt the brain functions and causes parkinsonism-like syndrome, which is initially apparent through apathy, anorexia, insomnia, extreme fatigability, somnolence, and a labile mood (Rudgalvyte et al.2016; Martinez et al. 2013; Farina et al. 2013; Gorojod et al. 2015). Mn exposure increases the progression of Parkinson's like symptoms tremors, abnormal movements, hypokinesia, speech disturbance, increased muscle tone, increased sweating and salivation (Tuschl et al. 2012; Quadri et al. 2012). However, brain was found pathologically different from PD in substancia nigra and basal ganglia (Shen and Dryhurst 1998). Epidemiological surveys on miners, workers on dry cell batteries and children's from buried dry cell batteries site have exhibited the neuropsychological behaviors, cognitive dysfunctions, emotional and motor defectiveness concomitant with an alteration (speedy) movement.

Arsenic is a major toxicant, found mostly in contaminated ground water which, makes an alloy with other metals (Prakash et al. 2016). Inorganic and methylated forms of arsenic accumulated near the different part of the brain and impair the normal functional mechanism of brain (Waly et al. 2016). Though, arsenic acts as teratogen, which may cross the placenta, and impairs the brain development through the formation of neural tube (Martinez et al. 2008). At the biochemical and molecular level, arsenic obstructs the cellular and molecular mechanism, where, an imbalance of Ca^{++}, mitochondrial dysfunction, and oxidative stress, disruption of ATP, altered membrane potential, cellular morphology, neuronal death and reduction on glial cells were reported (Yin et al. 1994). These biological changes in the brain leads to physiological impairment on vocabulary, mental acuity, language precision, total IQ and comprehending abilities such as difficulty in assembling the pictures with sequencing. Arsenic induces the beta amyloid formation which has been known for the main cause of AD (Giasson et al. 2002). It impairs the quality of life and metal health with a high level of depression, anxiety with psychiatric disorder and insomnia (Ashok et al. 2017). Arsenic play a major role as carcinogen if present in a small amount of drinking water and causes severe intestinal pain, vomiting, diarrhea, muscle cramps, cardiac arrhythmias (Jarup 2003).

Cadmium has been used in many industrial factories of batteries, electroplating, solder, nuclear reactor shield, cigarette smokers and dental amalgams. Most of the cadmium enters in the body by inhalation via the olfactory bulb, which interrupts the function of BBB (Evans and Hastings 1992). Cd exposure induce the function of the nervous system and generates neurological disorder (Wang and Du 2013). Cd is well known factor of AD and PD (Jiang et al. 2007; Okuda et al. 1997). Cd also accelerates the accumulation of the *Tau* proteins, which is responsible for AD (Jiang et al. 2007). It mostly impact the brain in comparison to other organs (Agnihotri et al. 2015) and develops the neurological dysfunctions like decrease in learning ability, headache, and olfactory dysfunction (Mason et al. 2014). Cd has been known for morphological changes in axons and dendrites of the brain, decrease in size and inhibits the neurite growth (Baker et al. 1983). Cd produces free radicals in the brain, which radically damages the neurons and oligodendrocytes (Parkinson et al. 1986). Oligodendrocyte used to form the myelin sheath around neurons, which conduct the nerve signal in the form of electrical impulses, obstructs the signal by Cd induced free radical damage. Cd directly abolishes choroid plexus structure, by which a barrier of CSF and spifflicate the filtration process of unsolicited materials (Pal et al. 1993) that defects the motor disabilities, learning inability, behavior defects, brain lesions and neurochemical changes. Effects of Cd toxicity, maximum in, cerebral cortical neurons of brain have been identified (Bishak et al. 2015), where oxidative stress induced by mitochondrial dysfunction disrupts the Ca^{++} ion signaling process and leads to apoptosis in primary murine neurons (Orrenius and Nicotera 1994a). Neurogenesis is also affected by Cd, which leads to less number of neuronal and glial cells (Gottofrey and Tjalve 1991; Chow et al. 2008). Apoptosis is a major concern at the time of development, proliferation and differentiation, where in some cases, proliferation reduced to 50%. Epigenetic effects of Cd is also known because of weak association with DNA, and methylation, which disrupts the whole gene functions (Zevin and Benowitz 1999). DNA methylation is the best study of the epigenetic process that regulates the gene silencing (Fig. 4).

Lead (Pb) is used in many industries of coloring material such as painting industry, hair coloring, batteries, cables, solder, electroplating and in petroleum industries. Lead (Pb) is absorbed by children in much more amount than adults due to under-developed BBB. It has been reported that even a very low level (250 ppm) of Pb can injure the hippocampus (Jett et al. 1997). Lead (Pb) can get accumulated in different regions of brain and mainly in the hippocampus (Jett et al. 1997). Lead (Pb) contact outcomes in discrepancy in language, memory, executing the task, verbal concept formation, poor reasoning, and poor command following (Hussien et al. 2018). Lead (Pb) is reported to increase the Tau protein phosphorylation in the cortex and cerebellum, which is a primary cause of AD and PD (Rai et al. 2010). Blockade of receptor on the membrane or disruption in the structure of membrane receptor by Pb influences the mechanism of neural plasticity, affects the long-term potentiation and memory loss (Rai et al. 2010; Baranowska-Bosiacka et al. 2012). Schizophrenia, autism, Attention-Deficit Hyperactivity Disorder (ADHD) has chance to exposure

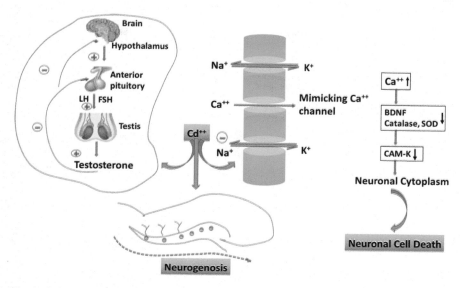

Fig. 4 Cadmium impaired the function of Na$^+$/K$^+$ ATPase enzyme and mimicking the function of Ca^{++} channel in neurons, which inhibited the expression of brain-derived neurotrophic factors (BDNF), catalase, superoxide dismutase (SOD) and increase the intracellular Ca^{++}, increased the free radicles which causes cell death. Cadmium has been detected as a reason to inhibit proliferation and differentiation in the hippocampus. Cadmium mimics the estrogen effects, in the reproductive system and disturbs the hypothalamic-pituitary-gonadal (HPG) axis

from environmental pollution from Pb (Modabbernia et al. 2016). With an increase of environmental pollutants, Pb toxicity has increased in humans, especially in children. However, if Pb toxicant reaches to the body in early childhood, may be disastrous for life, and produces abnormal behavior. This may results to brain toxicity and influence the learning and cognitive ability, memory impairment, and retrieval of memory, by affecting the normal life. Exposures to Pb may cause severe neurological diseases in children during childhood. Moreover, this may perturb the BBB, disrupts the NMDA receptor, phosphorylation of Tau/AB protein for AD and PD. This may also imbalance the Ca^{++} ion concentration, by increasing the PKC activity, and diminishing the BDNF level due to oxidative stress and which develops Autism, AD, PD, depression, anxiety, ALS (Zheng et al. 2003) (Fig. 5).

Mercury (Hg) as heavy metal, is widespread used in several industrial process. Mercury is mostly used in chemical research industry, laboratory purposes, medicine, electric industry, cosmetics, firearms and mercury lamp industry. Prevalent uses of mercury in industrial purposes, may affect the worker's life and its contamination in water and air persuades many biological irregularities. Inorganic Hg is incapable to cross the BBB but its alloy, methyalmercury (MeHg) cross the BBB and impairs the brain (Sheehan et al. 2014). Hg of aquatic arena mostly methylated and modified to MeHg due to presence of sulphate reducing bacteria (Parks et al. 2013). MeHg easily binds with aquatic and sea animals, therefore the consumption of these food chain

Fig. 5 High affinity of Pb^{++} towards NMDA receptor make complex, which inhibited to take glutamate, released from presynaptic neuron and NR2A (subunit of NMDA receptor) released the Ca^{++} in low amount, in postsynaptic neuron disrupts the cognitive dysfunction and memory impairment

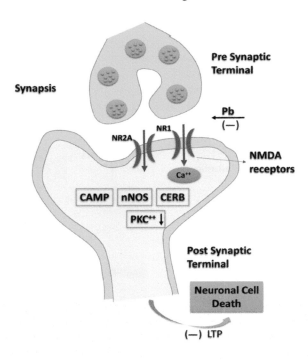

organisms by human remains fundamental form of exposure to Hg. High affinity of MeHg towards sulphur increases the binding to thiol group of protein and may cross the BBB (Suzuki et al. 1976). MeHg crosses the BBB as amino acid transporters and can bind to cysteine amino acid and mimic methionine (Takeda et al. 2000). It crosses the BBB and gets distributed in the different parts of brain like occipital lobe, basal ganglia and cerebellum (Davis et al. 1994), to impede the neuronal activities. This also includes dopamine secretion, neural stem cell differentiation, aberrant mitophagy, mitochondrial dysfunction, due to oxidative stress (Chang et al. 2018; Yuntao et al. 2016). Some population-based study has reveled mercury in the blood of children suffering from autism (Pamphlett and Kum Jew 2016), that risks children of 2–5 years of age.

Thallium (TI) is heavy metals found in the earth crust. Thallium has been used in several industries such as electronics, mercury lamps, jewelry, pigmentation, scintillation counters, and semiconductors for different purposes. Thallium is also used in cement and rodenticide industry, and from there, thallium contaminate the soil and gets exposed to human body (Galván-Arzate and Santamaría 1998; Cvjetko et al. 2010). Thallium enters into the body with skin contact or by inhalation of thallium contaminated air and consumption of food from contaminated soil or water. Thallium is reasoned behind numerous neurological diseases and non-neurological diseases. Non-neurological symptoms of thallium are anorexia, vomiting, gastrointestinal bleeding, abdominal pain, paresthesia, alopecia, cardio-toxicity with arrhythmias, and coma like life threating diseases occurs due to expoures. Thallium is also reasoned to cause seizure, fatigue, emotional changes, delirium, hallucination, ataxia, and loss of sensation, cranial-nerve deficit, and polyneuropathy like neurological symptoms in a dose dependent manner (Saha 2005; Zhao et al. 2008; Pelclová et al. 2009). Cortex, cerebellum and brain stem are the main regions of brain which are affected by thallium (Ríos et al. 1989). Moreover, toxicity was activated through lipid peroxidation and lysosomal enzyme beta-galectosydase in brain regions into dose dependent manner (Osorio-Rico et al. 2017). High concentration of thallium accumulated in hypothalamus and the potassium ion concentration may leads to many neurodegenerative diseases (Diaz and Monreal 1994).

Several metal in mixture of different concentration changes the homeostasis of biological environments of the body but have not been studied very extensively and minutely. Pb, As, Cd, Hg metals are studied in animal models as a mixture to examine the toxicity level and their multiple implications on various mechanisms, which may lead to neurological disorder.

4 Neurotoxicity Mechanism Induced by Heavy Metals

Heavy metals are used in our daily life, where, minute quantity may cause major impact in homeostasis, tiny quantity affects the enzyme activity in the brain and an excess quantity may generate neurotoxicity, which leads to neurodegenerative disorder. Most of the neurological diseases are developed during late old age and the reasons are unknown. However, it can be assume that these heavy metals may changes the behavior of enzymatic activity, oxidation of cellular molecules, mitochondrial dysfunctions, and molecular or cellular activities in the minute level at the initial stage in the brain. Metal absorption is different in the brain from other organs, where, most of the metals absorbs in gastrointestinal tract, skin, lungs, olfactory organs, and mixed in the circulatory system. Metals reached inside the brain to cross the BBB from blood to CSF and CSF to the inner side of the brain (Lemercier et al. 2003). Metal diffused from blood, through BBB, to CSF. In some cases, metal disrupts the BBB, and make permeable to other pathogens, compounds to generate autoimmune and neurological diseases like meningitis, epilepsy, multiple sclerosis, de novo diseases, cerebral edema, brain trauma, an damyotrophic lateral sclerosis.

Metals have an affinity towards divalent metal ion-transporter-1 (DMT-1) and transferrin (Tf) receptors; interact with BBB and brain tissue (Yokel et al. 2006). Passing through the BBB, glial fibrillary acidic protein (GFAP) expression reduced. Metals effect and disrupts the BBB, and make permeable to pathogens to generate autoimmune neurodegenerative diseases like multiple sclerosis, epilepsy, amyotrophic lateral sclerosis, meningitis, cerebral edema, and systematic inflammation. Divalent cation transporter (DMT-1) and transferrin transporter also binds with heavy metals at BBB and takes to CSF, where they accumulate in different parts of the brain. Metals involves in the mechanisms to bind with NMDA receptors, which are important for the cognitive behaviors and memory. Generally, NMDA receptors responsible for glutamate balance in postsynaptic regions and bind to glutamate induced the efflux of Ca^{++} to neuronal cytoplasm of postsynaptic region (Farina et al. 2013). Mostly Pb causes irreversible inhibition of NMDA receptors by binding, and disrupts the Ca^{++} signaling in neuronal synapsis (Kim et al. 2005). This may also generates pathological neurobiological symptoms which has been accountable for memory function (Neal et al. 2011). Pb transformed the NMDA receptor subunits NR2A and NR2B ratio which perturb the Ca^{++} signaling in hippocampus and altered the BDNF, n-NOS, CERB, liable for memory and long-term potentiation inhibition (Guilarte et al. 2000). Calcium ion is an indispensable for neuron homeostasis, which maintains the intracellular signaling of presynaptic terminal, cell body, and dendritic arbor by balancing the Ca^{++} efflux (Rai et al. 2010). Pb causes the disruption of Ca^{++} concentration and activates the phosphor kinase C (PKC) and phosphor lipase C (PLC). This may unregulates the phosphor kinase A (PKA) and calmodulin dependent protein kinases (CMKII), followed to decreased the BDNF, CERB, antioxidant enzyme with increased ROS and MAPK signaling might leads the demise of neuronal cell (Reinholz et al. 1999). Pb affects the pre-syneptic (VAMP1/2, synaptophysin, synaptotagmin-1, SNAP25, syntaxin-1) and postsynaptic proteins (PSD-95), which may induce the pathological changes in the cerebellum, forebrain cortex and hippocampus. Moreover, synaptic region had swollen, elongated and shrunken mitochondria in Pb treated animals (Gassowska et al. 2016). These symptoms implies the synaptic dysfunction and disruption of neurotransmission mechanism due to Pb, and leads to neurodegenerative disorder and neurobiological diseases (Ashok et al. 2015).

Excessive accumulation of metals, found in different parts of the brain, may particiapte in the mechanism of oxidative damage to the brain cells by deformation of protein, accumulation of amyloid protein, disrupting the Na^+/K^+ ATPase pump, upset acetylcholine esterase (AchE), and imbalance of ubiquitous Ca^{++} ions (Kinoshita et al. 2014). The Cd induced oxidative stress and mitochondrial dysfunction were recognized by Kumar et al. (2013). Cd exposure increased the Ca^{++} concentration in the cytoplasm and nucleus of neurons and such changes may abandone the cellular functions (Orrenius and Nicotera 1994a, b). Ca^{++} signaling affects the mitochondrial dysfunction and generate excess amount of ROS in neurons. Therefore, it consequently reduce the defense mechanism by changing the levels of the anti-oxidant like, glutathione, catalase, and superoxide dismutase activity. Increased ROS level induce the apoptosis mechanism in neuronal cells by affecting caspase-3 and caspase-9 activity, resulting in neuronal cell death in various parts of the brain (Jung et al. 2008), which may lead to cause ALS, a neurodegenerative diseases (Hart and Gitler 2012). Cd reduce the mitochondrial membrane potential of mitochondria (Hossain et al. 2009), (a major source of energy for neurons). It also reduces the capacity of mitochondria to produce ATP by oxidative phosphorylation and interrupt the transportation of mitochondria in distal part of neurons (Han et al. 2017). Disruption in mitochondrial transportation implicated in several neurodegenerative diseases as Alzheimer, Huntington and Parkinson's diseases (Jiang et al. 2007). Abnormal Ca^{++} homeostasis and mitochondrial dysfunction also evokes ROS and active free radicals. These free radicals move to the nucleus for mutation in DNA of pre and post synaptic proteins. Increased level of ROS also reduces the BDNF level, which has vital role for memory and cognitive function in brain hippocampus (Baranowska-Bosiacka et al. 2012).

Hg proceeded as an organic form in humans, and its minute quantity has a major impact in the brain (Lohren et al. 2015). Hg found in mother milk also, which influences the brain development in new born (Johansson et al. 2007), resulting in cognitive dysfunction and disruption in learning behavior over the years. Hg interacts in cellular level and disrupts the function of neurotransmitter, microtubule and Ca^{++} homeostasis (Lafon Cazal et al. 1993). MeHg increased oxidative stress in cerebral cortex (Chang et al. 2013; Yuntao et al. 2016), causes the neuronal injury which induces amyotrophic lateral sclerosis/motor neuron diseases (ALS/MNS) (Chang et al. 2013; Yuntao et al. 2016). Thiol (-SH) group has high tendency to react with MeHg (Suzuki et al. 1976; Straka et al. 2016), present in cysteine and methionine amino acid of protein. MeHg impedes the thiol metabolism, which is a prime cause of autism (neurological disease of children at the age of 6), blamable for biochemical changes on transketolase, oxidative stress, abnormal thiamine homeostasis. Impeded thiol metabolism and oxidative stress decreased the glutathione (GSH), an anti-oxidant, present into high amount in the brain, and inhibited the activity of Na^+/K^+ ATPase, NADH dehydrogenase, glutathione reductase and oxidative stress, makes neural vulnerable for neurodegenerative diseases (Gibon et al. 2010). MeHg, is responsible for glutamate (Glu) toxicity in brain (Moretto et al. 2005), block uptake of glutamate in astrocytes, and further, glutamate amount increased in the extracellular fluid, causes toxicity in the spinal cord. Toxicity in spinal cord and high concentration of glutamate by MeHg affects the synaptic activity of neurons, leading to devastating neurodegenerative diseases (Moretto et al. 2005). Hg inhibits the activity of superoxide dismutase (SOD), glutathione peroxidases and elevates the ROS level, which may increase the oxidative stress in cerebral cortex. Moreover, NF-E2-related factor 2 (Nrf-2) pathway get activated (Yang et al. 2017) and Ca^{++} ATPase activity inhibited, by increased intracellular loading of Ca^{++} and unbalanced Ca^{++} signaling in neurons. This may inhibit the process of electro-chemical activity of neuron, which is essential for the information transfer between neurons in every part of the brain. Hg causes the disruption in mitochondrial membrane potential by abandoned Ca^{++} in mitochondria (LeBel et al. 1990) and inhibits the mitochondrial electron transport chain (ETC) in cultured neural cells. Mercury enhance nitric oxide production and the activates the glial cells in brain, which reduces the glutathione level in brain (Simmons-Willis et al. 2002). All these mechanisms induced by MeHg make the brain vulnerable for neurodegenerative diseases.

Exposure to heavy metals results into neurodegenerative diseases or neurobiological diseases with multiple mechanistic pathway like mitochondrial dysfunction, Ca^{++} signaling, ROS generation, apoptosis, autophagy, interrupted anti-oxidant enzymes activity and crucial signaling pathway. Arsenic (As) phosphorylates the *tau* protein, (a microtubule associated protein of neurons), where this phosphorylation makes aggregation of *tau* proteins, deregulate the function of tau protein and causes neurodegeneration (Alizadeh-Ghodsi et al. 2018). Arsenic also cause the inflammation and degeneration of neural cells by ROS production and inhibition of antioxidant activity in neural stem cells, reasoned for multiple sclerosis (MS) and impaired neural activity (Sun et al. 2017; Alizadeh-Ghodsi et al. 2018). Arsenic induces the inflammatory process in the cerebrum; cerebellum, thalamus and brain stem by modifying the inflammatory genes. Inflammation instigates the myelin reduction in nerve cells into the central nervous system, which hindered the electric signal in between neurons and cause neurodegenerative diseases and multiple sclerosis. Other inflammatory mechanism activates the microglial cells, cytokine interleukin-6 (IL-6), and IFN-γ (released from astrocyte), which causes the injury of neurons, inhibition of regeneration, death of neurons and oligodendrocyte implicated in neurodegenerative diseases as PD, AD, HD and traumatic brain injury on different parts of the brain (Alizadeh-Ghodsi et al. 2018; Sun et al. 2017; Ashok et al. 2015).

Mixed metal effects also analyzed in animal to disclose the severity of metal on neurodegenerative diseases. Al, Pb, Hg may cause the neurotoxicity with deregulation of Ca^{++} homeostasis in the brain microsomes (Andrade et al. 2017). Changes in the flux of Ca^{++} has been presented as indexing of heavy metal neurotoxicity in brain (Bostanci and Bagirici 2013). Al, Pb, Hg neurotoxicity implemented by IP3 mediated calcium release (Pb > Hg > Al) and inhibition of calcium uptake by microsomes in brain (Pb > HG > Al) (Pentyala et al. 2010). Cd and Pb have combined effects of Na^+/K^+ATPase, where Pb made the reaction more potent and imbalance the Na^+/K^+, Ca^{++} intracellular manner. Pb, Cd and Arsenic elicits ROS, and stimulate the signaling pathway of ERK, JNK, MEPK to induce neurotoxicity inside the brain and seeding of neurodegeneration diseases (Nori et al. 1996). Every metal has their different pathways to react and generate its effect, so the cumulative effect is always more detrimental than a single metal. Further study of mixed metal, Pb, Cd, MeHg, Arsenics acts individually differently, such as Pb binds to NMDA receptors, Cd inhibits the Na^+/K^+ ATPase pump, MeHg inhibits the glutamate uptake, where all are responsible for Ca^{++} deregulation in neurons (Ahlskog et al. 1995). This instigate the mechanism to ROS generation, inhibition of antioxidant, reduced BDNF, induced apoptosis, mitochondrial deregulation, and causing the neural cell death (Wang and Du 2013; Stackelberg et al. 2013).

5   Signaling Mechanism to Brain Cancer

Abnormal growth of a cell inside the brain causes a brain tumor. Brain tumor is very rare (2%) in human, and it has been found in two forms, benign and cancerous (or malignant). Cancerous form is divided in two, primary tumors (within a brain), and malignant (spread from other organs) (metastasis). Till now, the cause of cancer is unknown and only few mechanisms speculates the brain cancer, which found in glial cell (glioblastoma) and meningioma (benign) (Lathia et al. 2015). Glioblastoma can developed in astrocytes, oligodendrocytes and neural stem cells, which have metastatic activities, known as cancer stem cells (Omuro and DeAngelis 2013). Contrary occupational studies showed that metals have no role in brain cancer, as it has not been detected in brain cancer patients (Wesseling et al. 2002). Mechanism behind the glioblastoma development, further reflected same in heavy metal mechanism, as comparable mechanism thought to be reason behind the initiation of brain cancer. In glioblastoma, the mechanism of calcium ion imbalance or irregular calcium signaling, mitochondrial dysfunction, increase in apoptotic pathway, restriction in autophagy, irregular cell cycle, JNK, ERK, MAPK signaling, Na^+/K^+ ATPase plays very prominent role, which are reprise by metals. Pb, Cd, As and Hg may induce the calcium ion deregulation in neural cells (Vu et al. 2018), which disrupts the mitochondrial dysfunction (Gugnani et al. 2018), disrupts the energy source and make cells glucose dependent (founds mostly in glioblastoma cells) (Singh et al. 2005). Heavy metals induced ROS formation may initiate DNA damage, lipid peroxidation, disruption in protein activities, free radical generation and correlate with epigenetic of brain tumors (Zhang et al. 2017). Metals produced oxidative stress causes the DNA mutation, strand breakage, and DNA methylation, histone alteration. In heavy metal, arsenic hypermethylate the DNA in the promoter of CDKN2A, Ras associated domain family protein 1A and serine protease 3 (Cui et al. 2006). Explanatory study of mechanism induced by metal was obscure but some occupational studies reveals that metals presents in individual patients with brain cancer. There is elevated risk of low-grade glioblastoma in worker of metal industry (Van Wijngaarden and Dosemeci 2006). Pb also has carcinogenic activity (Arslan et al. 2011). Several other studies also indicated that Pb causes the high risk of cerebral tumor (Cocco et al. 1998, 1999).

6 Conclusions and Future Recommendation

Heavy metal, As, Pb, Cd, Hg, Al has been known for their toxic effects in all aquatic and mammalian species. Toxicity of these metals are susceptible for the brain and cause neurotoxicity, which impacts neurodegenerative diseases like, Alzheimer disease, Parkinson's disease, Amyotrophic lateral Sclerosis, Multiple sclerosis etc. Every individual metal has their own mechanism to cross the blood brain barrier and reach the brain; some make complexes with proteins and some transported by specific transporter receptors to CNS. Choroid plexus is also playing an important role into crossing the metal to CNS, from where they reach different brain parts and induce neurotoxicity, including changes in the memory and cognitive functions of the brain. Pb, Cd, Hg and Arsenic have a separate mechanisms in diverse part of the brain to affect neural activity such as Pb bind to the NMDA receptor and prevents taking glutamate, which affects cognitive disability, memory and LTP. Cd induces the ROS formation, mitochondrial dysfunctions, imbalance of Ca^{++}, oxidative stress in neuronal cells, causes the death of neurons in ALS diseases. Metal induced oxidative stress initiate the signaling process of ERK, JNK, and MEPK, which activate the glial cells to produce immunogenic response. Activated glial cells persuade the immune system, release cytokine IL-6, IFN-γ which may cause damage to neurons and degenerate neurons of the nervous system. There is not much evidences of metal in a sample of the brain cancer patients; however, the possibility is high in workers of metal industry to get neurological diseases with lowered level of glioblastoma.

It's evident from the studies that the lucid mechanism caused by the metal is obscure. Prolonged exposure of the metal and induction of brain cancer mechanism and neurodegenerative diseases at molecular level is completely ambiguous. Future research is represented as an innovation of molecular mechanism which might be able to explain the association of heavy metals and different types of brain tumors such as glioblastoma, meningitis.

References

Agnihotri SK, Agrawal U, Ghosh I (2015) Brain most susceptible to cadmium induced oxidative stress in mice. J Trace Elem Med Biol 30:184–193

Ahlskog JE, Waring SC, Kurland LT, Petersen RC, Moyer TP, Harmsen WS, Bush V (1995) Guamanian neurodegenerative disease: investigation of the calcium metabolism/heavy metal hypothesis. Neurology 45(7):1340–1344

Alizadeh-Ghodsi M, Zavvari A, Ebrahimi-Kalan A, Shiri-Shahsavar MR, Yousefi B (2018) The hypothetical roles of arsenic in multiple sclerosis by induction of inflammation and aggregation of tau protein: a commentary. Nutr Neurosci 21(2):92–96

Andrade VM, Aschner M, Marreilha Dos Santos AP (2017) Neurotoxicity of Metal Mixtures. Adv Neurobiol 18:227–265

Ankley GT, Bennett RS, Erickson RJ, Hoff DJ, Hornung MW, Johnson RD, Mount DR, Nichols JW, Russom CL, Schmieder PK, Serrrano JA, Tietge JE, Villeneuve DL (2010) Adverse outcome pathways: a conceptual framework to support ecotoxicology research and risk assessment. Environ Toxicol Chem 29(3):730–741

Arini A, Gourves PY, Gonzalez P, Baudrimont M (2015) Metal detoxification and gene expression regulation after a Cd and Zn contamination: an experimental study on Danio rerio. Chemosphere 128:125–133

Arslan M, Demir H, Arslan H, Gokalp AS, Demir C (2011) Trace elements, heavy metals and other biochemical parameters in malignant glioma patients. Asian Pac J Cancer Prev 12(2):447–451

Aschner M (2007) Manganese: recent advances in understanding its transport and neurotoxicity. Toxicol Appl Pharmacol 221:131–147

Aschner M, Clarkson TW (1989) Methyl mercury uptake across bovine brain capillary endothelial cells in vitro: the role of amino acids. Pharmacol Toxicol 64:293–297

Ashok A, Rai NK, Tripathi S, Bandyopadhyay S (2015) Exposure to As-, Cd-, and Pb-mixture induces Abeta, amyloidogenic APP processing and cognitive impairments via oxidative stress-dependent neuroinflammation in young rats. Toxicol Sci 143(1):64–80

Ashok BS, Ajith TA, Sivanesan S (2017) Hypoxia-inducible factors as neuroprotective agent in Alzheimer's disease. Clin Exp Pharmacol Physiol 44(3):327–334

Baker EL, Feldman RG, White RF (1983) The role of occupational lead exposure in the genesis of psychiatric and behavioral disturbances. Acta Psychiatr Scand Suppl 303:38–48

Baranowska-Bosiacka I, Gutowska I, Rybicka M, Nowacki P, Chlubek D (2012) Neurotoxicity of lead hypothetical molecular mechanisms of synaptic function disorders. Neurol Neurochir Pol 46(6):569–578

Bélanger M, Allaman I, Magistretti PJ (2011) Brain energy metabolism: focus on astrocyte-neuron metabolic cooperation. Cell Metab 14(6):724–738

Bishak YK, Payahoo L, Osatdrahimi A (2015) Mechanisms of cadmium carcinogenicity in the gastrointestinal tract. Asian Pac J Cancer Prev 16(1):9–21

Bostanci MÖ, Bagirici F (2013) Blocking of L-type calcium channels protects hippocampal and nigral neurons against iron neurotoxicity. The role of L-type calcium channels in iron-induced neurotoxicity. Int J Neurosci 123(12):876–882

Caffo M, Caruso G, Fata GL, Barresi V, Visalli M, Venza M, Venza I (2014) Heavy metals and epigenetic alterations in brain tumors. Curr Genomics 15(6):457–463

Caito S, Aschner M (2015) Neurotoxicity of metals. Handb Clin Neurol 131:169–189

Camandola S, Mattson MP (2017) Brain metabolism in health, aging, and neurodegeneration. EMBO J 36(11):1474–1492

Carocci A, Rovito N, Sinicropi MS, Genchi G (2014) Mercury toxicity and neurodegenerative effects. Rev Environ Contam Toxicol 229:1–18

Carpenter RL, Jiang BH (2013) Roles of EGFR, PI3 K, AKT, and mTOR in heavy metal-induced cancer. Curr Cancer Drug Targets 13(3):252–266

Caserta D, Graziano A, Monte G, Lo Bordi G, Moscarini M (2013) Heavy metals and placental fetal maternal barrier: a mini review on the major concerns. Eur Rev Med Pharmacol Sci 17(16):2198–2206

Chang SH, Lee HJ, Kang B (2013) Methylmercury induces caspase-dependent apoptosis and autophagy in human neural stem cells. J Toxicol Sci 38(6):823–831

Cherry JD, Olschowka JA, O'Banion MK (2014) Neuroinflammation and M2 microglia: the good, the bad, and the inflamed. J Neuroinflam 11:15

Chow ES, Hui MN, Lin CC, Cheng SH (2008) Cadmium inhibits neurogenesis in zebrafish embryonic brain development. Aquat Toxicol 87(3):157–169

Chowanadisai W, Kelleher SL, Lonnerdal B (2005) Zinc deficiency is associated with increased brain zinc import and LIV-1 expression and decreased ZnT-1 expression in neonatal rats. J Nutr 135:1002–1007

Christie KJ, Emery B, Denham M, Bujalka H, Cate HS, Turnley AM (2013) Transcriptional regulation and specification of neural stem cells. Adv Exp Med Biol 786:129–155

Clauberg M, Joshi JG (1993) Regulation of serine protease activity by aluminum: implications for Alzheimer disease. Proc Natl Acad Sci USA 90:1009–1012

Cocco P, Dosemeci M, Heineman EF (1998) Brain cancer and occupational exposure to lead. J Occup Environ Med 40(11):937–942

Cocco P, Heineman EF, Dosemeci M (1999) Occupational risk factors for cancer of the central nervous system (CNS) among US women. Am J Ind Med 36(1):70–74

Cui X, Wakai T, Shirai Y, Hatakeyama K, Hirano S (2006) Chronic oral exposure to inorganic arsenate interferes with methylation status of p16INK4a and RASSF1A and induces lung cancer in A/J mice. Toxicol Sci 91(2):372–381

Cvjetko P, Cvjetko I, Pavlica M (2010) Thallium toxicity in humans. Arh Hig Rada Toksikol 61(1):111–119

Davis LE, Kornfeld M, Mooney HS (1994) Methylmercury poisoning: long-term clinical, radiological, toxicological, and pathological studies of an affected family. Ann Neurol 35(6):680–688

Davson H, Welch K, Segal MB (1987) The Secretion of the Cerebrospinal Fluid. The physiology and pathophysiology of the cerebrospinal fluid. Churchill Livingstone, New York, pp 218–221

Desai V, Kaler SG (2008) Role of copper in human neurological disorders. Am J Clin Nutr 88(3):855S–858S

Diaz RS, Monreal J (1994) Thallium mediates a rapid chloride/hydroxyl ion exchange through myelin lipid bilayers. Mol Pharmacol 46(6):1210–1216

Doorn JM, Kruer MC (2013) Newly characterized forms of neurodegeneration with brain iron accumulation. Curr Neurol Neurosci Rep 13(12):413

Dorman DC, Brenneman KA, McElveen AM, Lynch SE, Roberts KC, Wong BA (2002) Olfactory transport: a direct route of delivery of inhaled manganese phosphate to the rat brain. J Toxicol Environ Health A 65:1493–1511

Dusek P, Litwin T, Czlonkowska A (2015) Wilson disease and other neurodegenerations with metal accumulations. Neurol Clin 33(1):175–204

Evans J, Hastings L (1992) Accumulation of Cd(II) in the CNS depending on the route of administration: intraperitoneal, intratracheal, or intranasal. Fundam Appl Toxicol 19(2):275–278

Farina M, Avila DS, Da Rocha JB (2013) Metals, oxidative stress and neurodegeneration: a focus on iron, manganese and mercury. Neurochem Int 62(5):575–594

Fondell E, O'Reilly EJ, Fitzgerald KC, Falcone GJ, McCullough ML, Park Y, Kolonel LN, Ascherio A (2013) Magnesium intake and risk of amyotrophic lateral sclerosis: results from five large cohort studies. Amyotroph Lateral Scler Frontotemporal Degener 14(5–6):356–361

Galván-Arzate S, Santamaría A (1998) Thallium toxicity. Toxicol Lett 99(1):1–13

Gassowska M, Baranowska-Bosiacka I, Moczydlowska J, Frontczak-Baniewicz M, Gewartowska M, Struzynska L, Adamczyk A (2016) Perinatal exposure to lead (Pb) induces ultrastructural and molecular alterations in synapses of rat offspring. Toxicology 373:13–29

Giacoppo S, Galuppo M, Calabrò RS, D'Aleo G, Marra A, Sessa E, Bua DG, Potortì AG, Dugo G, Bramanti P, Mazzon E (2014) Heavy metals and neurodegenerative diseases: an observational study. Biol Trace Elem Res 161(2):151–160

Giasson BI, Sampathu DM, Wilson CA, Vogelsberg-Ragaglia V, Mushynski WE, Lee VMY (2002) The environmental toxin arsenite induces tau hyperphosphorylation. Biochemistry 41(51):15376–15387

Gibon J, Tu P, Frazzini V, Sensi SL, Bouron A (2010) The thiol-modifying agent N-ethylmaleimide elevates the cytosolic concentration of free $Zn(2^+)$ but not of $Ca(2^+)$ in murine cortical neurons. Cell Calcium 48(1):37–43

Giralt M, Molinero A, Carrasco J, Hidalgo J (2000) Effect of dietary zinc deficiency on brain metallothionein-I and -III mRNA levels during stress and inflammation. Neurochem Int 36(6):555–562

Gorojod RM, Alaimo A, Porte Alcon S (2015) The autophagic-lysosomal pathway determines the fate of glial cells under manganese- induced oxidative stress conditions. Free Radic Biol Med 87:237–251

Gottofrey J, Tjalve H (1991) Axonal transport of cadmium in the olfactory nerve of the pike. Pharmacol Toxicol 69(4):242–252

Graves AB, Rosner D, Echeverria D, Mortimer JA, Larson EB (1998) Occupational exposures to solvents and aluminium and estimated risk of Alzheimer's disease. Occup Environ Med 55:627–633

Greenough MA, Camakaris J, Bush AI (2013) Metal dyshomeostasis and oxidative stress in Alzheimer's disease. Neurochem Int 62(5):540–555

Greter M, Merad M (2013) Regulation of microglia development and homeostasis. Glia 61:121–127

Grubman A, Pollari E, Duncan C, Caragounis A, Blom T, Volitakis I, Kanninen KM (2014) Deregulation of biometal homeostasis: the missing link for neuronal ceroid lipofuscinoses? Metallomics 6(4):932–943

Gugnani KS, Vu N, Rondon-Ortiz AN, Bohlke M, Maher TJ, Pino-Figueroa AJ (2018) Neuroprotective activity of macamides on manganese-induced mitochondrial disruption in U-87 MG glioblastoma cells. Toxicol Appl Pharmacol 340:67–76

Guilarte TR, McGlothan JL, Nihei MK (2000) Hippocampal expression of N-methyl-d-aspartate receptor (NMDAR1) subunit splice variant mRNA is altered by developmental exposure to Pb2+. Mol Brain Res 76(2):299–305

Halliwell B, Gutteridge JMC (2007) Free radicals in biology and medicine, 4th edn. Oxford University Press

Han J, Yang X, Chen X, Li Z, Fang M, Bai B, Tan D (2017) Hydrogen sulfide may attenuate methylmercury-induced neurotoxicity via mitochondrial preservation. Chem Biol Interact 263:66–73

Harris WR (1983) Thermodynamic binding constants of the zinchuman serum transferrin complex. Biochemistry 22:3920–3926

Hart MP, Gitler AD (2012) ALS-associated ataxin 2 polyQ expansions enhance stress-induced caspase 3 activation and increase TDP-43 pathological modifications. J Neurosci 32(27):9133–9142

Heppner FL, Ransohoff RM, Becher B (2015) Immune attack: the role of inflammation in Alzheimer disease. Nat Rev Neurosci 16:358–372

Hosovski E, Mastelica Z, Sunderic D, Radulovic D (1990) Mental abilities of workers exposed to aluminium. Med Lav 81:119–123

Hossain S, Liu HN, Nguyen M, Shore G, Almazan G (2009) Cadmium exposure induces mitochondria-dependent apoptosis in oligodendrocytes. Neurotoxicology 30(4):544–554

Hussien HM, Abd-Elmegied A, Ghareeb DA, Hafez HS, Ahmed HEA, El-Moneam NA (2018) Neuroprotective effect of berberine against environmental heavy metals-induced neurotoxicity and Alzheimer's-like disease in rats. Food Chem Toxicol 111:432–444

Hyder F, Patel AB, Gjedde A, Rothman DL, Behar KL, Shulman RG (2006) Neuronal-glial glucose oxidation and glutamatergic-GABAergic function. J Cereb Blood Flow Metab 26:865–877

Jarup L (2003) Hazards of heavy metal contamination. Br Med Bull 68:167–182

Jett DA, Kuhlmann AC, Farmer SJ, Guilarte TR (1997) Age dependent effects of developmental lead exposure on performance in the morris water maze. Pharmacol Biochem Behav 57(1–2):271–279

Jiang LF, Yao TM, Zhu ZL (2007) Impacts of Cd(II) on the conformation and self-aggregation of Alzheimer's tau fragment corresponding to the third repeat of microtubule-binding domain. Biochim Biophys Acta 1774(11):1414–1421

Joana S, Cristóvão RS, Cláudio MG (2016) Metals and neuronal metal binding proteins implicated in Alzheimer's Disease. Oxid Med Cell Long 17–47

Johansson C, Castoldi AF, Onishchenko N, Manzo L, Vahter M, Ceccatelli S (2007) Neurobehavioural and molecular changes induced by methylmercury exposure during development. Neurotox Res 11(3–4):241–260

Jung YS, Jeong EM, Park EK, Kim YM, Sohn S, Lee SH, Moon CH (2008) Cadmium induces apoptotic cell death through p38 MAPK in brain microvessel endothelial cells. Eur J Pharmacol 578(1):11–18

Kawahara M, Muramoto K, Kobayashi K, Mori H, Kuroda Y (1994) Aluminum promotes the aggregation of Alzheimer's amyloid beta-protein in vitro. Biochim Biophys Res Commun 198:531–535

Killin LO, Starr JM, Shiue IJ, Russ TC (2016) Environmental risk factors for dementia: a systematic review. BMC Geriatr 16(1):175

Kim MJ, Dunah AW, Wang YT, Sheng M (2005) Differential roles of NR2A- and NR2B-containing NMDA receptors in Ras-ERK signaling and AMPA receptor trafficking. Neuron 46(5):745–760

Kinoshita PF, Yshii LM, Vasconcelos AR, Orellana AM, Lima Lde S, Davel AP, Scavone C (2014) Signaling function of Na, K-ATPase induced by ouabain against LPS as an inflammation model in hippocampus. J Neuroinflam 11:218

Koppenal C, Finefrock AE, Bush AI, Doraiswamy PM (2004) Copper, iron and zinc as therapeutic targets in Alzheimer's disease. Res Prac Alzheimer's Dis 9:250–255

Kumar P, Sannadi S, Reddy R (2013) Alterations in apoptotic caspases and antioxidant enzymes in arsenic exposed rat brain regions: reversal effect of essential metals and a chelating agent 6(3):1150–1166

Kumudini N, Uma A, Devi YP, Naushad SM, Mridula R, Borgohain R, Kutala VK (2014) Association of Parkinson's disease with altered serum levels of lead and transition metals among South Indian subjects. Indian J Biochem Biophys 51(2):121–126

Lafon Cazal M, Pietri S, Culcasi M, Bockaert J (1993) NMDA dependent superoxide production and neurotoxicity. Nature 364(3):535–537

Lathia JD, Mack SC, Mulkearns-Hubert EE, Valentim CL, Rich JN (2015) Cancer stem cells in glioblastoma. Genes Dev 29(12):1203–1217

LeBel CP, Ali SF, McKee M, Bondy SC (1990) Organometal induced increases in oxygen reactive species: the potential of 2′,7′-dichlorofluorescin diacetate as an index of neurotoxic damage. Toxicol Appl Pharmacol 104(1):17–24

Lemercier V, Millot X, Ansoborlo E, Menetrier F, Flury-Herard A, Rousselle C, Scherrmann JM (2003) Study of uranium transfer across the blood-brain barrier. Radiat Prot Dosimetry 105(1–4):243–245

Li W, Garringer HJ, Goodwin CB, Richine B, Acton A, VanDuyn N, Vidal R (2015) Systemic and cerebral iron homeostasis in ferritin knock-out mice. PLoS One 10(1):e0117435

Liu Y, Wu H, Kou L, Liu X, Zhang J, Guo Y, Ma E, (2014) Two metallothionein genes in Oxya chinensis: molecular characteristics, expression patterns and roles in heavy metal stress. PLoS One 12:9(11)

Loef M, Walach H (2015) Iron and copper in alzheimer's disease: a review micronutrients in dementia and cognitive decline, pp 563–571

Lohren H, Blagojevic L, Fitkau R, Ebert F, Schildknecht S, Leist M, Schwerdtle T (2015) Toxicity of organic and inorganic mercury species in differentiated human neurons and human astrocytes. J Trace Elem Med Biol 32:200–208

Lopez GC, Varea E, Palop JJ, Nacher J, Ramirez C, Ponsoda X, Molowny A (2002) Cytochemical techniques for zinc and heavy metals localization in nerve cells. Microsc Res Tech 56(5):318–331

Lynam DR, Roos JW, Pfeifer GD, Fort BF, Pullin TG (1999) Environmental effects and exposures to manganese from use of methylcyclopentadienyl manganese tricarbonyl (MMT) in gasoline. Neurotoxicol 20:145–150

Mantyh PW, Ghilardi JR, Rogers S, DeMaster E, Allen CJ, Stimson ER, Maggio JE (1993) Aluminum, iron, and zinc ions promote aggregation of physiological concentrations of beta-amyloid peptide. J Neurochem 61:1171–1174

Martinez EJ, Kolb BL, Bell A, Savage DD, Allan AM (2008) Moderate perinatal arsenic exposure alters neuroendocrine markers associated with depression and increases depressive-like behaviors in adult mouse offspring. Neurotoxicology 29(4):647–655

Martinez-Finley EJ, Gavin CE, Aschner M (2013) Manganese neurotoxicity and the role of reactive oxygen species. Free Radic Biol Med 62:65–75

Mason LH, Harp JP, Han DY (2014) Pb neurotoxicity: neuropsychological effects of lead toxicity. Biomed Res Int 2014:840547

McCarthy RC, Sosa JC, Gardeck AM, Baez AS, Lee CH, Resnick MW (2018) Inflammation-induced iron transport and metabolism by brain microglia. J Biol Chem RA118.001949. https://doi.org/10.1074/jbc.ra118.001949

Mergenthaler P, Lindauer U, Dienel GA, Meisel A (2013) Sugar for the brain: the role of glucose in physiological and pathological brain function. Trends Neurosci 36(10):587–597

Meyer E, Kurian MA, Hayflick SJ (2015) Neurodegeneration with brain iron accumulation: genetic diversity and pathophysiological mechanisms. Annu Rev Genomics Hum Genet 16:257–279

Modabbernia A, Velthorst E, Gennings C, De Haan L, Austin C, Sutterland A, Reichenberg A (2016) Early-life metal exposure and schizophrenia: a proof-of-concept study using novel tooth-matrix biomarkers. Eur Psychiatry 36:1–6

Mohamed Fel B, Zaky EA, El-Sayed AB, Elhossieny RM, Zahra SS, Salah Eldin W, Youssef WY, Khaled RA, Youssef AM (2015) Assessment of hair aluminum, lead, and mercury in a sample of autistic Egyptian children: environmental risk factors of heavy metals in Autism. Behav Neurol 2015:545674

Mohandas B, Colvin RA (2004) The role of zinc in Alzheimer's disease. Recent Res Devel Physiol 2:225–245

Moretto MB, Funchal C, Santos AQ, Gottfried C, Boff B, Zeni G, Rocha JB (2005) Ebselen protects glutamate uptake inhibition caused by methyl mercury but does not by Hg2+. Toxicology 214(1–2):57–66

Murakami M, Hirano T (2008) Intracellular zinc homeostasis and zinc signaling. Cancer Sci 99(8):1515–1522

Nathanson D, Mischel PS (2011) Charting the course across the blood-brain barrier. J Clin Invest 121(1):31–33

Neal AP, Worley PF, Guilarte TR (2011) Lead exposure during synaptogenesis alters NMDA receptor targeting via NMDA receptor inhibition. Neurotoxicology 32(2):281–289

Nelson N (1999) Metal ion transporters and homeostasis. EMBO J 18:4361–4371

Nori A, Fulceri R, Gamberucci A, Benedetti A, Volpe P (1996) Biochemical and functional heterogeneity of rat cerebrum microsomal membranes in relation to SERCA $Ca(2 +)$-ATPases and $Ca2 +$ release channels. Cell Calcium 19(5):375–381

Notarachille G, Arnesano F, Calo V, Meleleo D (2014) Heavy metals toxicity: effect of cadmium ions on amyloid beta protein 1-42. Possible implications for Alzheimer's disease. Biometals 27(2):371–388

Okuda B, Iwamoto Y, Tachibana H (1997) Parkinsonism after acute cadmium poisoning. Clin Neurol Neurosurg 99(4):263–265

Oliver KR, Fazakerley JK (1998) Transneuronal spread of Semliki Forest virus in the developing mouse olfactory system is determined by neuronal maturity. Neurosci 82:867–877

Omuro A, DeAngelis LM (2013) Glioblastoma and other malignant gliomas: a clinical review. JAMA. 310(17):1842–1850

Orrenius S, Nicotera P (1994a) The calcium ion and cell death. J Neural Transm Suppl 43:1–11

Orrenius S, Nicotera P (1994b) The calcium ion and cell death. J Neural Transm Suppl 43:1–11

Osorio-Rico L, Santamaria A, Galvan-Arzate S (2017) Thallium toxicity: general issues, neurological symptoms, and neurotoxic mechanisms. Adv Neurobiol 18:345–353

Pal R, Nath R, Gill KD (1993) Influence of ethanol on cadmium accumulation and its impact on lipid peroxidation and membrane bound functional enzymes (Na + , K + -ATPase and acetylcholinesterase) in various regions of adult rat brain. Neurochem Int 23(5):451–458

Pamphlett R, Kum Jew S (2016) Locus ceruleus neurons in people with autism contain no histochemically-detectable mercury. Biometals 29(1):171–175

Parkinson DK, Ryan C, Bromet EJ (1986) A psychiatric epidemiologic study of occupational lead exposure. Am J Epidemiol 123(2):261–269

Parks JM, Johs A, Podar M (2013) The genetic basis for bacterial mercury methylation. Science 339(6125):1332–1335

Pelclová D, Urban P, Ridzon P (2009) Two-year follow-up of two patients after severe thallium intoxication. Hum Exp Toxicol 28(5):263–272

Pentyala S, Ruggeri J, Veerraju A, Yu Z, Bhatia A, Desaiah D, Vig P (2010) Microsomal Ca2 + flux modulation as an indicator of heavy metal toxicity. Indian J Exp Biol 48(7):737–743

Peres TV, Schettinger MR, Chen P, Carvalho F, Avila DS, Bowman AB, Aschner M (2016) Manganese-induced neurotoxicity: a review of its behavioral consequences and neuroprotective strategies. BMC Pharmacol Toxicol 17(1):57

Perl DP (2001) The association of aluminum and neurofibrillary degeneration in Alzheimer's disease, a personal perspective. In Exley C (ed) Aluminum and Alzheimer's disease. Elsevier, pp 133–146

Prakash C, Soni M, Kumar V (2016) Mitochondrial oxidative stress and dysfunction in arsenic neurotoxicity: A review. J Appl Toxicol 36(2):179–188

Quadri M, Federico A, Zhao T (2012) Mutations in SLC30A10 cause parkinsonism and dystonia with hypermanganesemia, polycythemia, and chronic liver disease. Am J Hum Genet 90(3):467–477

Rai A, Maurya SK, Khare P, Srivastava A, Bandyopadhyay S (2010) Characterization of developmental neurotoxicity of As, Cd, and Pb mixture: synergistic action of metal mixture in glial and neuronal functions. Toxicol Sci 118(2):586–601

Rao DB, Wong BA, McManus BE, McElveen AM, James AR, Dorman DC (2003) Inhaled iron, unlike manganese, is not transported to the rat brain via the olfactory pathway. Toxicol Appl Pharmacol 193:116–126

Reinholz MM, Bertics PJ, Miletic V (1999) Chronic exposure to lead acetate affects the development of protein kinase C activity and the distribution of the PKCgamma isozyme in the rat hippocampus. Neurotoxicology 20(4):609–617

Riihimaki V, Hanninen H, Akila R, Kovala T, Kuosma E, Paakkulainen H, Valkonen S, Engstrom B (2000) Body burden of aluminum in relation to central nervous system function among metal inert-gas welders. Scand J Work Environ Health 26:118–130

Ríos C, Galván-Arzate S, Tapia R (1989) Brain regional thallium distribution in rats acutely intoxicated with Tl2SO4. Arch Toxicol 63(1):34–37

Robison G, Sullivan B, Cannon JR (2015) Identifcation of dopaminergic neurons of the substantia nigra pars compacta as a target of manganese accumulation. Metallomics 7(5):748–755

Roos PM, Vesterberg O, Nordberg M (2006) Metals in motor neuron diseases. Exp Biol Med (Maywood) 231(9):1481–1487

Rudgalvyte M, Peltonen J, Lakso M (2016) RNA-seq reveals acute manganese exposure increases endoplasmic reticulum related and lipocalin mRNAs in caenorhabditis elegans. J Biochem Mol Toxicol 30(2):97–105

Saha A (2005) Thallium toxicity: a growing concern. Indian J Occup Environ Med 9(2):53–56

Salvador GA, Uranga RM, Giusto NM (2010) Iron and mechanisms of neurotoxicity. Int J Alzheimers Dis 2011:720658

Serlin Y, Shelef I, Knyazer B, Friedman A (2015) Anatomy and physiology of the blood-brain barrier. Semin Cell Dev Biol 38:2–6

Shaw CA, Tomljenovic L (2013) Aluminum in the central nervous system (CNS): toxicity in humans and animals, vaccine adjuvants, and autoimmunity. Immunol Res 56(2–3):304–316

Sheehan MC, Burke TA, Navas-Acien A (2014) Global methylmercury exposure from seafood consumption and risk of developmental neurotoxicity: a systematic review. Bull World Health Organ 92(4):254–269F

Shen XM, Dryhurst G (1998) Iron- and manganese-catalyzed autoxidation of dopamine in the presence of l-cysteine: possible insights into iron- and manganesemediated dopaminergic neurotoxicity. Chem Res Toxicol 11(7):824–837

Simmons-Willis TA, Koh AS, Clarkson TW, Ballatori N (2002) Transport of a neurotoxicant by molecular mimicry: the methylmercury-L-cysteine complex is a substrate for human L-type large neutral amino acid transporter (LAT) 1 and LAT2. Biochem J 367(Pt 1):239–246

Singh D, Banerji AK, Dwarakanath BS, Tripathi RP, Gupta JP, Mathew TL, Jain V (2005) Optimizing cancer radiotherapy with 2-deoxy-d-glucose dose escalation studies in patients with glioblastoma multiforme. Strahlenther Onkol 181(8):507–514

Sparks DL, Schreurs BG (2003) Trace amounts of copper in water induce beta-amyloid plaques and learning deficits in a rabbit model of Alzheimer's disease. Proc Natl Acad Sci USA 100:11065–11069

Stackelberg KV, Elizabeth G, Tian C, Birgit CH (2013) Mixtures, metals, genes and pathways: a systematic review. Working Paper prepared for: methods for research synthesis: a cross-disciplinary workshop. Harvard Center for Risk Analysis, 3 Oct 2013

Straka E, Ellinger I, Balthasar C, Scheinast M, Schatz J, Szattler T, Gundacker C (2016) Mercury toxicokinetics of the healthy human term placenta involve amino acid transporters and ABC transporters. Toxicology 340:34–42

Sun X, He Y, Guo Y, Li S, Zhao H, Wang Y, Xing M (2017) Arsenic affects inflammatory cytokine expression in Gallus gallus brain tissues. BMC Vet Res 13(1):157

Suzuki T, Shishido S, Ishihara N (1976) Different behaviour of inorganic and organic mercury in renal excretion with reference to effects of D-penicillamine. Br J Ind Med 33(2):88–91

Szabo ST, Harry GJ, Hayden KM, Szabo DT, Birnbaum L (2016) Comparison of Metal Levels between Postmortem Brain and Ventricular Fluid in Alzheimer's Disease and Nondemented Elderly Controls. Toxicol Sci 150(2):292–300

Takeda A, Suzuki M, Okada S, Oku N (2000) Zn localization in rat brain after intracerebroventricular injection of 65Zn-histidine. Brain Res 863:241–244

Tuschl K, Clayton PT, Gospe SM Jr (2012) Syndrome of hepatic cirrhosis, dystonia, polycythemia, and hypermanganesemia caused by mutations in SLC30A10, a manganese transporter in man. Am J Hum Genet 90(3):457–466

Tykwinska K, Lauster R, Knaus P, Rosowski M (2013) Growth and differentiation factor 3 induces expression of genes related to differentiation in a model of cancer stem cells and protects them from retinoic acid-induced apoptosis. PLoS One 8(8):e70612

Van Wijngaarden E, Dosemeci M (2006) Brain cancer mortality and potential occupational exposure to lead: findings from the national longitudinal mortality study, 1979–1989. Int J Cancer 119:1136–1144

Vu HT, Kobayashi M, Hegazy AM, Tadokoro Y, Ueno M, Kasahara A, Hirao A (2018) Autophagy inhibition synergizes with calcium mobilization to achieve efficient therapy of malignant gliomas. Cancer Sci. (https://doi.org/10.1111/cas.13695 (Epub ahead of print)

Waly M, Power-Charnitsky VA, Hodgson N, Sharma A, Audhya T, Zhang Y, Deth R (2016) Alternatively spliced methionine synthase in SH-SY5Y neuroblastoma cells: cobalamin and GSH dependence and inhibitory effects of neurotoxic metals and thimerosal. Oxid Med Cell Longev 6143753

Wang B, Du Y (2013) Cadmium and its neurotoxic effects. Oxid Med Cell Longev 2013:898034

Wegst USR, Mullin EJ, Ding D, Manohar S, Salvi R, Aga DS, Roth JA (2015) Endogenous concentrations of biologically relevant metals in rat brain and cochlea determined by inductively coupled plasma mass spectrometry. Biometals 28(1):187–196

Wesseling C, Pukkala E, Neuvonen K, Kauppinen T, Boffetta P, Partanen T (2002) Cancer of the brain and nervous system and occupational exposures in Finnish women. J Occup Environ Med 4(7):663–668

White DM, Longstreth WT, JrL Rosenstock, Claypoole KH, Brodkin CA, Townes BD (1992) Neurologic syndrome in 25 workers from an aluminum smelting plant. Arch Intern Med 152:1443–1448

Wu MJ, Hu HH, Siao CZ, Liao YM, Chen JH, Li MY, Chen YF (2018) All Organic Label-like Copper(II) Ions Fluorescent Film Sensors with High Sensitivity and Stretchability. ACS Sens 3(1):99–105

Yang H, Wu J, Guo R, Peng Y, Zheng W, Liu D, Song Z (2013) Glycolysis in energy metabolism during seizures. Neural Regen Res 8(14):1316–1326

Yang T, Xu Z, Liu W, Feng S, Li H, Guo M, Xu B (2017) Alpha-lipoic acid reduces methylmercury-induced neuronal injury in rat cerebral cortex via antioxidation pathways. Environ Toxicol 32(3):931–943

Yassa HA (2014) Autism: a form of lead and mercury toxicity. Environ Toxicol Pharmacol 38(3):1016–1024

Yin JC, Wallach JS, Del Vecchio M, Wilder EL, Zhou H, Quinn WG, Tully T (1994) Induction of a dominant negative CREB transgene specifically blocks long term memory in drosophila. Cell 79(1):49–58

Yokel RA (2006) Blood brain barrier flux of aluminum, manganese, iron and other metals suspected to contribute to metal induced neurodegeneration. J Alzheimers Dis 10(2–3):223–253

Yokel RA, Florence RL (2006) Aluminum bioavailability from the approved food additive leavening agent acidic sodium aluminum phosphate, incorporated into a baked good, is lower than from water. Toxicology 227(1–2):86–93

Yuntao F, Chenjia G, Panpan Z (2016) Role of autophagy in methylmercury-induced neurotoxicity in rat primary astrocytes. Arch Toxicol 90(2):333–345

Zeineh MM, Chen Y, Kitzler HH, Hammond R, Vogel H, Rutt BK (2015) Activated iron-containing microglia in the human hippocampus identified by magnetic resonance imaging in Alzheimer disease. Neurobiol Aging 36:2483–2500

Zevin S, Benowitz NL (1999) Drug interactions with tobacco smoking. An update. Clin Pharmacokinet 36(6):425–438

Zhang Z, Miah M, Culbreth M, Aschner M (2016) Autophagy in neurodegenerative diseases and metal neurotoxicity. Neurochem Res 41(1–2):409–422

Zhang C, Jiang H, Wang P, Liu H, Sun X (2017) Transcription factor NF-kappa B represses ANT1 transcription and leads to mitochondrial dysfunctions. Sci Rep 7:44708

Zhao G, Ding M, Zhang B (2008) Clinical manifestations and management of acute thallium poisoning. Eur Neurol 60(6):292–297

Zheng W, Aschner M, Ghersi-Egea JF (2003) Brain barrier systems: a new frontier in metal neurotoxicological research. Toxicol Appl Pharmacol 192(1):1–11

Zucca FA, Segura-Aguilar J, Ferrari E, Muñoz P, Paris I, Sulzer D, Sarna T, Casella L, Zecca L (2017) Interactions of iron, dopamine and neuromelanin pathways in brain aging and Parkinson's disease. Prog Neurobiol 155:96–119

Molecular Mechanisms of Heavy Metal Toxicity in Cancer Progression

Pragati Singh, Deepak Tiwari, Manish Mishra and Dhruv Kumar

Abstract In last few years, cancer became one of the leading cause of death in humans. There are several factors associated with the cancer initiation and progression including heavy metals. Several heavy metals including arsenic, cadmium, uranium, lead, mercury etc. and heavy metal-containing compounds are toxic to the humans and have been reported to induce mutations in human genome which further leads to the carcinogenesis. This chapter provides the detail understanding of molecular mechanisms and pathway analysis to heavy metal toxicity in human carcinogenesis.

Keywords Heavy metals · Cancer · Arsenic · Cadmium · Cobalt · Uranium

1 Introduction

Heavy metal is simple collaboration of two entirely different words i.e. heavy and metal or we can say the metals which are heavy. The definition says that density of metal is high but actually this physical quantity is quite useless in the cases of plants and other living organisms (Jaishankar et al. 2014). Plants and living organisms do not deal with metals or we can say that these are not accessible to them in their elemental form (Jaishankar et al. 2014).

If we want to define these two words then heavy means something that have weight and metals means substances or elements which can conduct heat and electricity and have properties like ductility, malleability and luster (Aziz et al. 2008; Florea et al.

P. Singh · D. Tiwari · D. Kumar (✉)
Amity Institute of Molecular Medicine & Stem Cell Research (AIMMSCR), Amity University
Uttar Pradesh, J3-112, Sec-125, Noida, India
e-mail: dkumar13@amity.edu; dhruvbhu@gmail.com

P. Singh · D. Tiwari
Centre of Bioinformatics, University of Allahabad, Allahabad 211002, Uttar Pradesh, India

M. Mishra
Health Physics Division, Bhabha Atomic Research Centre, Trombay, Mumbai, India

© Springer Nature Switzerland AG 2019
K. K. Kesari (ed.), *Networking of Mutagens in Environmental Toxicology*, Environmental Science,
https://doi.org/10.1007/978-3-319-96511-6_3

49

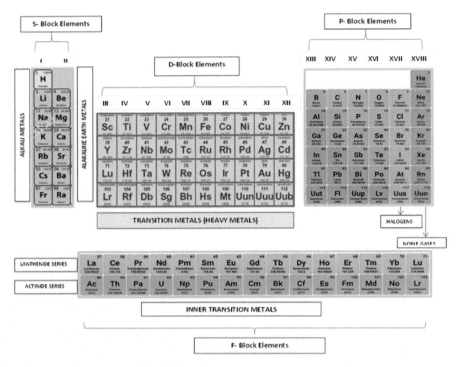

Fig. 1 Positions of heavy metals in periodic table

2012). The metals having a property i.e. temperature dependent conductivity which differentiate it from non-metals and metalloids (Jaishankar et al. 2014). Therefore heavy metals can simply be defined as elements those have a density i.e. 5 times greater than the specific density of water. The specific gravity of water is found to be 1 at a temperature of 4 °C (39 °F) (Ilyin et al. 2004). Well, we can describe specific density as a ratio in between the density of any substances to the density of some other substances considered as standard under specified conditions of pressure and temperature.

In case of liquids and solids we can consider water as standard and in case of gases hydrogen or air is considered as a standard (Aziz et al. 2008; Brochin et al. 2008). It is symbolized as sp.gr. This quantity is dimensionless and therefore not expressed in units. The heavy metals are chemical components or elements that are mostly found in the earth crust (Järup 2003; Mamtani et al. 2011. There are a specific place and description of each and every heavy metal in the periodic table. It is believed that there must be a correlation in between "toxicity" and "heaviness" (Fig. 1).

Human beings are blessed with power to understand things in a far better way than other living organisms. Humans, instead of using this power as an asset they are deteriorating and destroying the environment in which they live. Modernization has involved the use of toxic metals more than its limit value which is being causing a variety of health hazards (Kim et al. 2015).

There are basically two types of Health Hazards: (1) Hazards associated with most target organ effects, (2) Hazards associated with cancer and mutagenic effects. Inappropriate and over exploitation of our resources has a big hand in problems caused due to heavy metals (Kim et al. 2015). Various heavy metals like cadmium, mercury, arsenic, lead, etc. are being recklessly used in manufacture of items we use in our daily lives. In addition to this, the food we eat insecticides which contains Arsenic, mercury etc. chemicals in high amounts (Kawada 2016). Cadmium, a heavy metal is known to cause endometrial cancer (Mazariegos et al. 2010). Almost every heavy metal can cause cancer. Among these heavy metals few are needed by our body for good metabolism and possess various other functions in our body. Reactive oxygen species (ROS) causes oxidative stress that is proven to be a reason behind most of diseases caused by heavy metals (Grimsrud and Peto 2006). Cadmium, nickel, chromium and arsenic fall in 1st category according to International Agency for Research on Cancer (Su et al. 2007). Several reports show that vulnerability to these heavy metals causes interference in tumor suppressor gene expression, ruins repair processes and enzymatic agitated in metabolism by oxidative harm (Zhang et al. 2007). Screening our soil with these harmful heavy metals can contaminate the vegetation and can result in oral cancer. The revolution of heavy metals in our bio-system results in concentration of high amount of toxic metal (Wen et al. 2005). Today presence of tremendous amount of biological data is an outcome of the increased attention towards heavy metal and its carcinogenetic impact on health. So the data mining is very essential method and can be counted into our major concern (Lee et al. 2016). Chemical/Gene peculiar pathways that are actually very complex can be understood with the help of the pathways studio databases as it dispense drawings for the pathways by using data that are collected from variety of sources (Yuan et al. 2011). We can use and analyze the pathways because it can give a better and more comparative view on carcinogenesis, diseases and marker proteins that are related with heavy metals (Khatri et al. 2012). Further, more coordinate network between marker proteins and cell forms adds to the expectation of carcinogenesis particular protein markers. Damages/Harm or hazards that are instigated by metal could be prevented and detoxified by engaging different inter-cellular chelation procedures and cancer prevention or anti-oxidants (Lobo et al. 2010). Resistance against metal poisoning is developed by combing metal ions with the phytochelatins (phytochelatins known as the chelating agents in plants). Oxidative damages are overcome by the complex formed when molecules of anti-oxidants interacts with free radicals ("and dusts," n.d.). Utilization of phytochemicals from cell reinforcement substances from plants can aid the cell reinforcement related detoxification processes.

2 Why It Is Important to Discuss Metal Toxicity?

Heavy metals can be further categorized into beneficial heavy metals and toxic heavy metals. Here, we lay our focus on the toxic heavy metals since they are of much concern for us (Hodson 2004; Jan et al. 2015; Patil et al. 2013). Out of 35 metals that are needed to be concerned, 23 are heavy metals small amounts of these metals are essential for good health but large intake is hazardous and can cause acute or chronic toxicity. Excessive intake causes damage to mental and central nervous function, lower energy levels, damaged blood composition, lungs, kidneys, liver and other vital organs (Arif et al. 2016; Singh et al. 2011). Long-term exposure to such metals can result in-slow progression in cancer, physical muscular and neurological degenerative processes, Alzheimer's disease, Parkinson's disease, muscular dystrophy, multiple sclerosis etc.

3   Heavy Metals not Associated with Cancer

As earlier stated heavy in small quantities are very essential and beneficial for health. They exist in various compounds and unimolecular forms and in various food stuffs.
   They have various medical and industrial applications

1. Indirect injection of gallium in the radiological procedure
2. In x-ray equipment 'lead' is used as a radiation shield.
3. In manufacturing pesticides, batteries, alloys, electroplated metal parts, textile dyes etc.
4. They also constitute important role in products used in our homes.

4   Toxic Heavy Metals

Heavy metals when not metabolized act as toxic heavy metals resulting in their accumulation in the soft tissues (Goyer et al. 2004). Their intake/pathways into the human body are through food, water, air, or adsorption by the skin in case of pharmaceutical or industrial setting (Barra et al. 2006; Mamtani et al. 2011. Their intake varies to different age groups such that adults develop metal toxicity generally through Industrial Exposure whereas in children it develops by eating toxic metals (Prüss-Ustün et al. 2011; Village 2005) in form of some substance or also from hand to mouth activity of small children when they touch dirty and contaminated soils etc.

   Other common routes of exposure to toxicity are during a radiological procedure, from inappropriate doing during parental nutrition, from parental nutrition, from broken thermometers. People residing in older homes with lead painted or old plumbing develop contamination (Length 2007; Morin et al. 2007; OEHHA 2001). Inhalation or skin contact with dust, fumes or vapor causes acute poisoning (Kazemipour et al. 2008; Sankhla et al. 2016). The agency for toxic substances and disease registry in Atlanta, Georgia in cooperation with U. S. Environmental Protection Agency has compiled a priority list named "The top 20 hazardous substances" in 2001. The heavy metals 1 Arsenic, 2 Lead, 3 Mercury, 7 Cadmium occur in this list.

Table 1 Role of Arsenic in cancer initiation and progression

Sources	Cancers associated with arsenic	Regulatory limitations	Target organs	Symptoms
Smelting	Lung cancer	Environmental Protection Agency (EPA)—0.01 parts per million (ppm) in drinking water	Blood	Sore throat
Manufacturing of chemicals and glasses	Bladder cancer	Occupational Safety and Health Administration (OSHA)—10 $\mu g/m^3$ of workplace air for 8 h shifts and 40 h work weeks	Kidneys	Reddening at the contact point
Production of arsenic gas released from pesticides containing arsenic	Skin cancer		Central nervous system	Severe abdominal pain
			Digestive system	Vomiting and diarrhea
			Skin system	Other symptoms include fever, mucosal irritation

4.1 Arsenic

Chronic exposure to Arsenic can result in damage to central nervous system, excessive skin darkening (hyperpigmentation) in areas not exposed to sunlight, excessive formation of skin on soles and palms (hyperkeratosis), or white bands of arsenic deposits across the fingernails (visible after 4–5 weeks of exposure) (Tchounwou et al. 2004). Cardiovascular changes are often diagnosed earlier which later can lead to cardiovascular collapse (Chioma et al. 2017; Soignet et al. 2001). Table 1 summarizing the sources of arsenic and associated factors.

4.2 Lead

Chronic exposure to lead leads to birth defects, mental retardation, autism, psychosis, allergies, dyslexia, hyperactivity, weight loss, shaky hands and muscular weakness (Stohs and Bagchi 1995; Tchounwou et al. 2012a, b). Children are often sensitive to lead toxicity. Symptoms of chronic lead exposures include—colic, allergies, autism, hyperactivity, mood swings, nausea, numbness, and lack of concentration (Kaul et al. 1999; Yedjou et al. 2010). Table 2 summarizng the sources of lead and associated factors.

Table 2 Role of Lead in cancer initiation and progression

Sources	Cancer associated with lead	Regulatory limitations	Target organs	Symptoms
Old houses (in painted surfaces)	Lung cancer	EPA—15 parts per billion (ppb) in drinking water, 0.15 μg/m³ in air	Bones	Abdominal pain
Industries	Stomach cancer		Brain	Convulsions
Fertilizers			Blood	Hypertension
			Kidneys	Loss of appetite, fatigue, and sleeplessness
			Thyroid gland	Renal dysfunction
				Other symptoms include- hallucination, headache, numbness, arthritis, and vertigo

4.3 Mercury

Mercury exists in three forms: Elemental mercury, organic and inorganic mercury.

Liquid mercury is more likely to be ingested by the children because of its beautiful colors and unique behavior when spilled. Common sources of liquid mercury are broken thermometer, or drinking medicine containing mercury (Rooney 2007). Chronic exposure to mercury can cause permanent problems to central nervous system and kidneys. Mercury is capable of entering into the placenta and accumulation of which can cause, mental retardation, brain damage, blindness, cerebral palsy, inability to speak. Table 3 summarizng the sources of mercury and associated factors.

4.4 Cadmium

Chronic exposure to cadmium causes hazardous disease resulting in chronic obstructive lung diseases renal disease and fragile bones. It has been concluded by various research studies especially performed on Egyptian females that in case of breast cancer the levels of cadmium and copper were found to be increased and levels of iron was reduced. Table 4 summarizng the sources of cadmium and associated factors.

Table 3 Role of Mercury in cancer initiation and progression

Sources	Cancer associated with mercury	Regulatory limitations	Target organs	Symptoms
Dental amalgam	Prostate cancer	EPA—2 parts per billion parts (ppb) in drinking water	Brain	Sore throat
Thermometers		FDA—1 part of methylmercury in a million parts of sea food	Kidney	Shortness of breath
Algaecides and childhood vaccines		OSHA—0.1 mg of organic mercury per cubic meter of workplace air and 0.05 mg/m^3 of metallic mercury vapor for 8-h shifts and 40-h work week		Metallic taste in the mouth
Fertilizers				Abdominal pain
Mining industry				Nausea
				Vomiting and diarrhea
				Headaches
				Weakness
				Visual disturbances
				Tachycardia
				Hypertension

4.5 Aluminum

Aluminum studies show that long-term exposure to aluminum could be a reason for developing Alzheimer's disease since the Alzheimer's patients were found with the significant amount of aluminum in their brain tissues (Al-fartusie and Mohssan 2017). Aluminum-based coagulants are also being used for water purification which has balanced its potential health concerns to much extent. Table 5 summarizng the sources of aluminum and associated factors.

4.6 Iron

Chronic overdose of iron causes its deposit in the heart which may cause death due to myocardial siderosis. Table 6 summarizng the sources of iron and associated factors. The high amount of ingestion of iron causes cellular toxicity and impaired oxidative

Table 4 Role of Cadmium in cancer initiation and progression

Sources	Cancer associated with cadmium	Regulatory limitations	Target organs	Symptoms
Nickel–cadmium batteries	Breast cancer	Primary drinking water standard (MCL) 0.005 mg/l	Liver	Nausea
PVC plastics	Lung cancer	Hazardous waste screening criteria (TCLP) 20 mg/kg	Placenta	Vomiting
Fertilizers		Livestock water quality 0.5 mg/l	Kidney	Abdominal pain
Reservoirs containing shellfish			Lungs	Breathing difficulty
Cigarettes			Brain	Learning disorders
Dental alloys			Bones	Loss of taste
Motor oil				Growth impairment
				Cardiovascular disease

phosphorylation and mitochondrial dysfunction resulting in cellular death (Griswold and Martin 2009).

4.7 Copper

Chronic exposure of copper causes damage to liver and kidney and destroys RBC's. Acute (short term) effects of copper causes temporary gastrointestinal distress. Though copper is essential for body as it helps fight anemia and necessary for normal metabolic functions in Human. Deficiency of copper causes low numbers of white blood cells, osteoporosis in infants and children, and defects in connective tissue leading to skeletal problems. Table 7 summarizng the sources of copper and associated factors.

Table 5 Role of Aluminum in cancer initiation and progression

Sources	Cancer associated with aluminum	Regulatory limitations	Target organs	Symptoms
Food additives	Breast cancer	Secondary drinking water standard 0.05–0.20 mg/l	Central nervous system	Memory loss
Antacids		Common range in soils 10,000–300,000 mg/kg	Kidney	Learning difficulty
Buffered aspirin		Livestock water quality 5.0 mg/l	Digestive system	Loss of coordination
Astringents				Disorientation
Nasal sprays				Mental confusion
Antiperspirants				Colic heartburn
From drinking water				Flatulence
Automobile exhaust				Headaches
Tobacco smoke				
Aluminum foil				
Storage containers				

Table 6 Role of Iron in cancer initiation and progression

Sources	Cancer associated with iron	Regulatory limitations	Target organs	Symptoms
Iron tablets	Lung cancer	Secondary drinking water standard 0.3 mg/l	Liver	Nausea
Drinking water	Colon cancer	Common range in soils 7,000–550,000 mg/kg	Cardiovascular system	Vomiting
Iron pipes	Bladder cancer		Kidney	Abdominal pain
Cookware				Hematemesis
				Diarrhea
				Significant fluid and blood loss

Table 7 Role of Copper in cancer initiation and progression

Sources	Cancer associated with copper	Regulatory limitations	Target organs	Symptoms
Mines	Breast cancer	Primary drinking water standard (MCL) 1.39 mg/l	Liver	Nausea
Wires	Ovarian cancer	Secondary drinking water standard 1.0 mg/l	Digestive system	Vomiting
Pipes	Lung cancer	Common range in soils 2–100 mg/kg		Abdominal pain
Sheet metal				

Table 8 Role of Nickel in cancer initiation and progression

Sources	Cancer associated with nickel	Regulatory limitations	Target organs	Symptoms
Alloys	Breast cancer	Common range in soils 5–500 mg/kg	Lungs	Chest pain
Jewelries		Livestock water quality 1.0 mg/l	Respiratory tract	Cough
Food items		Land application of sewage sludge 420 ppm	Kidney	Dyspnoea
Chocolate			Cardiovascular system	Dizziness

4.8 Nickel

Chronic Exposure to nickel causes lung cancer, nose cancer, larynx cancer and prostate cancer. Exposure to nickel and compounds containing nickel causes dermatitis known as "nickel itch" to sensitive people (Zofkova et al. 2017). Nickel is a micronutrient and essential for health. Possible sources for exposure of nickel to human are air we breathe, drinking water, eating food (vegetables contain nickel) or smoking cigarettes. Inhalation of nickel gas causes respiratory failure, lung embolism and chronic bronchitis. Table 8 summarizng the sources of nickel and associated factors.

4.9 Tin

Large amount of consumption of inorganic tin causes stomachaches, anemia, liver and kidney problems. Chronic poisoning to tin can result in neurological problems. Acute exposure can cause skin and eye irritation, respiratory irritation and gastrointestinal effect. High amount ingestion can be fatal. Table 9 summarizng the sources of tin and associated factors.

Table 9 Role of Tin in cancer initation and progression

Sources	Cancer associated with tin	Regulatory limitations	Target organs	Symptoms
Alloys	Prostate cancer	All food in solid form 230 ppm	Nervous system	Stomach aches
Soil	Testicular cancer	All food in liquid form 230 ppm	Hematological system	Eye irritation
Anti-foiling paint			Respiratory system	Respiratory irritation
			Lungs	

4.10 Uranium

Uranium (U) is the heaviest radioactive metal which occurs naturally in the environment. It has an atomic number, A, of 92. It is one of the actinide series elements which is well documented, extensively studied and highly explored by human beings. Its physical properties are given in Table 10. Natural uranium (Nat U) is a mildly radioactive element consists of three (radio) isotopes namely U-238, U-235 and U-234, which exist in almost a secular equilibrium with each other. It emanates mainly Alpha particles, beta particles and gamma rays. The most relevant isotopes and their half-lives are provided in Table 11 and the level of contamination of uranium in ground water in many countries given in Table 12. It is known that Alpha particles, which carry massive energy of 4–8 meV (Mega electron volts), pose almost no external hazard i.e. when it is present outside of the body. That is because its range in air is only a few centimeters and it can be stopped even by a piece of sheet. Alpha cannot even penetrate the dead layer of our skin. However, when ingested through water or food can cause of a lot of internal damage to the soft tissues.

United States Environmental Protection Agency (USEPA) has classified uranium as a confirmed human carcinogen (Group A) and has published guidelines to enforcement agencies to follow a Zero tolerance for its presence in drinking water, with a maximum permissible limit of 30 ppb. World Health Organization (WHO) has published a set of reports in which it has emphasized that limits of uranium in drinking water should be less than 15 ppb, which it later in 2011 changed to 30 ppb. In India Atomic Energy Regulatory Board (AERB) has fixed maximum permissible limits of 60 ppb (AERB 2004).

Since uranium it is a naturally occurring radionuclide with an estimated half-life of millions of year, it is present in varying proportions in Earth's crust, seawater, surface waters, groundwater, plants and animals. While its concentration is reported around 3 ppb (part per billion or μg/L) in different water bodies, including seawater, it occurs in Earth's crust at an average worldwide level of around 3 ppm (Hu and Gao 2008). As it is a naturally occurring and ubiquitously present radioactive element, it ingestion from food, drinking water and inadvertently from soil is a regular phenomenon. It has

Table 10 Physical properties of uranium metal

Density (high purity)	19.05 ± 0.02 g/cm^3
Density (industrial uranium)	18.85 ± 0.20 g/cm^3
Melting point	1.132 ± 1 °C
Boiling point	38,113 C
Heat of fusion	4.7 kcal/mole
Vapor pressure	10^{-4} mm
Thermal conductivity	0.071 cal/cm-s-°C
Electrical resistivity @ (25 °C)	35×106 Ω/cm^3
Mean coefficient of linear thermal expansion (random orientation)	16×10^{-6}/°C
Specific heat	6.65
Enthalpy (25 °C)	1,520 cal/mole
Entropy	12.0 cal/mole/°C

Table 11 Relevant radioisotopes of uranium

Radioisotope	Abundance (w/w %)	Radioactivity percent (%)	Half-life (years)
U-238	99.27	48.7	4.47×10^9
U-235	0.720	2.72	7.04×10^8
U-234	0.005	49.03	2.4×10^5
U-233	Trace, anthropogenic	–	1.6×10^5
U-232	Anthropogenic	–	68.9

Table 12 Occurrence of elevated levels of uranium in groundwater around the world (BARC REPORT)

S. No.	Location details	Uranium content (μg/L or ppb)
1	Finland	0.04–12,000
2	France	0.35–74.4
3	Germany	0.03–48
4	Switzerland	0–80
5	China	0.01–56
6	USA	0.01–652
7	Sweden	<2–470
8	Norway	<0.02–170
9	Jordan	0.04–1,400
10	Punjab, India	<0.2–644

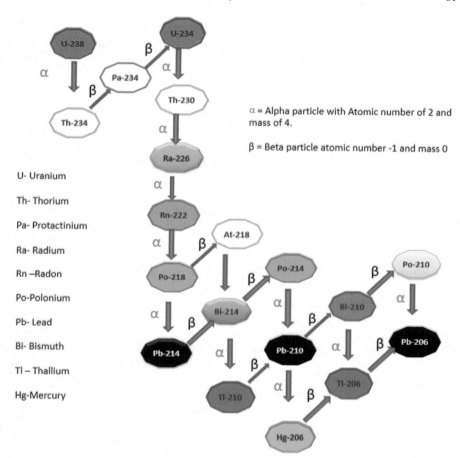

α = Alpha particle with Atomic number of 2 and mass of 4.

β = Beta particle atomic number -1 and mass 0

U- Uranium

Th- Thorium

Pa- Protactinium

Ra- Radium

Rn –Radon

Po-Polonium

Pb- Lead

Bi- Bismuth

Tl – Thallium

Hg-Mercury

Fig. 2 Decay series of uranium (U-238)

been estimated that an average adult human ingests (intakes) 1–2 μg/day of uranium from food and 1.5 μg/day from drinking water (Konietzka 2015; Singh et al. 1990). One gram of natural uranium having this relative isotopic abundance has an activity of 0.69 μCi (or 25,530 Bq as 1 Ci = 3.7 × 1,010 Bq). Of this 49.0% of the activity is attributable to U-234, 2.27% of the activity is attributable to U-235, and 48.7% of the activity is attributable to U-238 (Fig. 2).

Uranium when inhaled or ingested through air or food or water gets rapidly eliminated from the body. The maximum absorption in case of inhalation is 1–5% through lungs and in case of ingestion it is around 0.5–5%. Majority of inhaled uranium gets cleared from the lungs via mucociliary action or through fecal excretion of the swallowed sputum, however a very small portion of uranium may reside in the lungs for years. Ingested uranium (only 0.5–5%), on the other hand, gets absorbed in the blood from where it gets distributed initially to soft tissues and lymph nodes and then finally to kidney, liver and bone. Overall, most ingested uranium is excreted in feces and remainder in urine (Radespiel-Tröger and Meyer 2013).

Uranium, being an actinide series element can exist in $+II$, $+III$, $+IV$, $+V$ and $+VI$ oxidation states, however oxidation states of $+IV$ and $+VI$ are the ones which form a range of complex compounds in the environment and the only states which are abundant enough to study. The U(VI) oxidation states are mainly water-soluble compounds while $+IV$ are otherwise and are abundantly found in soils and rocks. U(IV) is rather insoluble and exists in complex forms with inorganic ligands e.g. fluoride, chloride, sulphate and phosphate. U(VI), as uranyl ($UO2++$) complex is abundant in wet soils and water mediums (Gómez et al. 2006).

In 20th century which lead to discovery of fission of uranium, to produce vast amount of energy trapped in the nucleus, has led to slight increase in worldwide concentration of fallout uranium. Extensive uranium mining, atmospheric nuclear tests, nuclear fuel recycling, use of depleted uranium in armours and waste disposal are some of the anthropogenic sources which cause increase in Uranium content of the environment accessible to human beings.

Apart from its presence in food, uranium gets inside human body through its presence in groundwater which is major source of drinking water worldwide. Slightly elevated levels of uranium in has been reported from all over the world including India. The reasons for elevated levels were previously attributed to leaching from soil due to excessive use of fertilizers or from fly-ashes produced from Thermal Power Plants operating on coal, which may be a case in few locations (Bajwa et al. 2017; Brindha and Elango 2013; Efstathiou et al. 2014; Liesch et al. 2015). However, recent developments and studies have found the source of uranium may be more of geogenic origin than anthropogenic (Liesch et al. 2015).

One of the peculiar phenomenon which occurs during the decay series of U-238 is production of a volatile gaseous daughter products, Radon-222 (Rn-222), which a half-life of 3.82 days. Rn-222 then diffuses out from the rocks, soil or aquifer sources to the atmosphere, dissolves in groundwater or sometime to crevices present in the nearby vicinity and the decays to its subsequent daughter products (Gokhale 2008). Although the half-life is in days, it is always found present in the environment as it is being continuously produced due to decay of natural uranium. Rn-222 is an alpha emitter with an energy of 5.5 meV and it is one the major cause of lung cancer in non-smokers. Rn-222 causes thousands of deaths worldwide because of its inhalation from milling, mining and cement and concrete materials which have elevated levels of natural uranium in it (Samet 2011). International Atomic Energy Agency (IAEA) has given a threshold limit for annual activity concentration of Rn-222 as 1,000 Bq/m^3 for building with high occupancy factor (IAEA 2015).

5 Methodologies for Radiological and Chemotoxic Risk Assessment of Uranium

Human health effects due to exposure of uranium can be classified as radiological risk (ionizing radiation effects of uranium isotopes) and as chemotoxic risk being a heavy metal. The radiological risk factor can be evaluated based on the general USEPA standard method (Hartmann et al. 2000). Using the risk factor and uranium level in subsurface water, the excess cancer risk which an average individual faces due to presence of uranium in drinking water can be calculated from below given equations (Kumar et al. 2011).

$$\textbf{Risk factor (per Bq/L)} = \text{Risk coefficient}$$
$$\times \text{Water Ingestion Rate} \times \text{total exposure duration} \quad (1)$$

where,

Risk coefficient (RC)	1.19×10^{-9} Bq,
Water Ingestion Rate (WIR)	4.05 L/day,
Total Exposure Duration	Avg. Life Expectancy (India, 63.7 years) $\times$ 365 = 23,250 days.

$$\textbf{Excess Cancer Risk} = \text{Uranium Concentration in groundwater} \times \text{Risk factor} \quad (2)$$

where,

Uranium concentration (Bq/L) Measured value (μg/L) $\times$ conversion factor (0.025 Bq/μg).

The chemotoxic risk can similarly be calculated based on the hazard quotient (HQ) and chemical toxicity risk in the form of Lifetime Average Daily Dose (LADD) were calculated through ingestion of groundwater by the following formula

$$\textbf{Hazard Quotient (HQ)} = \text{LADD/RfD} \quad (3)$$

$$\textbf{LADD} (\mu\text{g/kg/day}) = [\text{Ci} \times \text{IR} \times \text{EF} \times \text{LE}]/[\text{BW} \times \text{AT}] \quad (4)$$

where,

Ci	Concentration of U in subsurface water (μg/L),
IR	Ingestion rate (L/day),
EF	Exposure frequency (days/year),
LE	Life expectancy (years),
AT	Average Time (days),
BW	Bodyweight (kg),
RfD	Reference Dose (μg/kg/day),
LADD	Lifetime average daily dose, (μg/kg/day).

6   Clinical Effects of Various Toxic Metals

Arsenic: Arsenic causes perforation of nasal septum, respiratory cancer, peripheral neuropathy, dermatomes and skin cancer.

Cadmium: Cadmium   causes   proteinuria,   glucosuria,   osteomalacia, aminoaciduria, emphysemia.

Chromium: Chromium causes ulcer, perforation of nasal septum, respiratory cancer.

Manganese: Central and peripheral neuropathies.

Lead: Lead causes encephalopathy, peripheral neuropathy, Central nervous disorders, and anemia.

Nickel: Nickel causes cancer and dramatis.

Tin: Tin causes central nervous system disorders, visual defects and EEG changes and Pneumoconiosis.

Mercury: Mercury causes proteinuria.

7   Mechanism of Toxicity and Carcinogenicity of Some Specific Heavy Metals

7.1   Chromium

Main factors that determine the toxicity of chromium compounds are oxidation state and solubility. Chromium (VI) compounds are considered more toxic and irritation and corrosion. They are also better oxidizing agents (Mamtani et al. 2011; Dayan and Paine 2001). Inspite of the fact that biological mechanisms are not known, but the level of toxicity of various states of chromium can be explained as the more easily Cr(VI) can pass through cell membranes and further intracellular reduction to reactive intermediates (Adenocarcinoma et al. 2014). Therefore Cr(VI) is more toxic than Cr(III) i.e. poorly absorbed by any route.

Cr(VI) to Cr(III) extracellularly can help reducing in toxic effects of chromium. Cr(VI) form can be soaked up by the gastrointestinal tract, lungs and even up to certain extent by skin (Chioma et al. 2017; Odewabi and Ekor 2017; Stohs and Bagchi 1995). Cr(VI) can be reduced under physiological conditions by hydrogen peroxide (H_2O_2), glutathione (GSH) reductase, ascorbic acid and GSH to form reactive intermediates which includes Cr(V), Cr(IV), thiylradicals, hydroxyl radicals, and ultimately, Cr(III). Any of these species can attack/strike DNA, proteins, membrane lipids henceforth in disturbing cellular integrity and function (O'Brien et al. 2003).

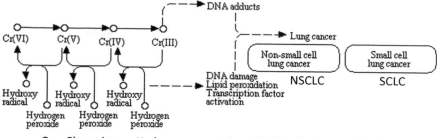

Cr = Chromium   Hydrogen peroxide = H2O2   Hydroxy radical = OH

Fig. 3 The figure is obtained from KEGG databases showing mechanism of Carcinogenesis induced by chromium metal. The oxidised and reduced forms of chromium metal act as a genotoxic carcinogens which leads to the formation of a DNA adduct and non genotoxic carcinogens which activate a transcription factor that causes DNA damage respectively. Further altogether these processes results into NSCLC and SCLC

7.1.1 Carcinogenesis Due to Chromium

According a report of epidemiological investigations workers are found suffering from respiratory cancer due to exposure to Cr(VI) containing compounds in their occupational environment (Clarkson 1993; O'Brien et al. 2003; Tchounwou et al. 2012a, b). Oxidative damage is considered as hidden cause of the genotoxic effects which include chromosomal abnormalities and DNA stand breaks (Dayan and Paine 2001) (Fig. 3). However recent studies show a biological relevance of non-oxidative mechanisms in Cr(VI) carcinogenesis. Carcinogenicity seems to be linked with inhalation of less soluble ore insoluble Cr(VI) compounds. Cr(VI) isn't toxic in its elemental form. Toxicity shows a vast variation to different Cr(VI) compounds (Clarkson 1993; Dayan and Paine 2001). Epidemiological evidence state Cr(VI) as a factor in Carcinogenesis (Goulart et al. 2005).

Solubility and other properties of chromium such as size, crystal modification, surface charge and the ability to be phagocytized could be significant in determining cancer risk (IARC 2006; Yamashoji and Isshiki 2001). Hypothetical concepts have been proposed to explain the carcinogenicity of chromium and its salts, although there have some issues from initial when discussing metal carcinogenesis because its different compounds have different potencies. Due to exposure of multiple chemicals in the industries thus it would be hard to conclude the carcinogenicity from any single metal (Browning et al. 2014; Duffus 2002). Hence carcinogenic risk is said to be caused not because of any single metal. Thus, carcinogenic risk often can be said to cause not because of any single but due to some group of metals.

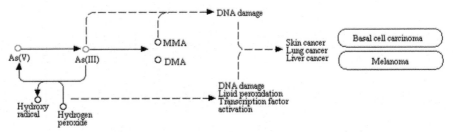

As = Arsenic   MMA= Monomethylarsonic acid   DMA = Dimethylarsinic acid

Fig. 4 The obtained figure is from KEGG database showing mechanism of carcinogenesis induced by arsenic metal. The reduction and oxidation phenomena of arsenic metal is responsible for the activation of lipid transcription factor and formation of two products i.e. monomethylarsonic acid and dimethylarsinic acid respectively that causes DNA damage and finally leads to the occurrence of different types of skin, lung and liver cancers

7.2 Arsenic

Solubility, oxidation state as well as several other extrinsic and intrinsic parameters strongly affects the toxicity. In case of Arsenic toxicity these factors in addition with many other factors that are reported by several research conducts lie frequency and time-period, exposure dose, gender and age, the biological species, along with genetic and nutritional parameters and person susceptibilities play a key role in regulating the toxicity levels (Puccetti et al. 2000). Exposure to inorganic As (arsenic) can be considered as a reason for a large number of human toxicity caused by the Arsenic.

In comparison to pentavalent arsenate As(V), the inorganic trivalent arsenite As(III) possess 2–10 times more toxicity (Hong et al. 2014; Velma and Tchounwou 2011). The As(III) attacks on the sulfhydryl groups or thiol of the protein and form a complex with them. By this way it can stop the activity of more than 200 enzymes (Jhaa et al. 1992). This mechanism is mainly responsible for the arsenic toxicity, effects of which can widely observed on various organ systems (Dong 2002; Mumtaz et al. 2002; Walker et al. 2010). Whereas pentavalent arsenate As(V) induces an exchange of phosphate group which in implicated in various biochemical pathways (Basu et al. 2001; Length 2007; Mamtani et al. 2011. The uncoupling of oxidative phosphorylation and the suppression of many mitochondrial enzymes causes impairment of cellular respiration through this mechanism arsenic impart its toxic effects (Fig. 4).

Enzymes like thiolase and dihydrolipoyl dehydrogenase becomes inactive when sulfhydryl groups of protein react with arsenic as result of which the processes of oxidation and beta oxidation of pyruvate and fatty acids respectively gets inhibited. In humans methylation is main metabolic pathway for inorganic arsenic (Basu et al. 2001; Sankhla et al. 2016; Tchounwou et al. 2004). Through a Non-enzymatic process, the Arsenic trioxide gets methylated to two main metabolites. Firstly it gets converted into monomethylarsonic acid (MMA) and before the discharge in the

Urine it again enzymatically methylated to dimethyl arsenic acid (DMA) (Chioma et al. 2017; Soignet et al. 2001). According to recent research it has been found that methylater metabolites can be more hazardous in comparison to arsenite if they possess arsenic trivalent forms (Stevens et al. 2010; Takahashi et al. 2002). Arsenic compounds have ability that they can restrict the process of DNA repair, can bring out chromosomal aberrations, replacement in between sister-chromatid and also causes organization of micronuclei in both rodent cells in culture and humans and in exposed human cells (Khoury et al. 2015; Liu et al. 1996; Odewabi and Ekor 2017; Verin et al. 1998).

7.3 Cadmium

Cadmium is an extreme pneumonic and gastrointestinal aggregation, which can be lethal if breathed in or ingested (Chioma et al. 2017; Zhang et al. 2004). Side-effects like muscle cramps, vertigo, stomach pain, burning sensations, spewing, sickness (nausea), lack of consciousness, shock and convulsions are typically observed within 15–30 min if taken in small amounts. It's consumption or intake in small amount can also lead to problem like disintegration of gastrointestinal tract, hepatic or renal, pneumatic damage and chronic unconsciousness i.e. coma and is totally based upon the course (routes) of poisoning (Kippler et al. 2012; Odewabi and Ekor 2017). A negative Impact has been observed on the serotonin, acetylcholine and norepinephrine levels upon the persistent exposure to chromium.

Pulmonary adenocarcinomas can be caused by chronic inhalation of cadmium and it is proved by the experiments or research works conducted upon the rodents (Mumtaz et al. 2002; Skipper et al. 2016; Yedjou and Tchounwou 2007). Systematic or direct subjection can also be a factor for the prostatic proliferate lesions that in turn contain adenocarcinomas. In spite of the fact that here we have an inadequate information about the mechanism of cadmium toxicity it has been observed that the reason of cell destruction is mainly the production of ROS, which in turns leads to destruction of the single stranded DNA and distorted synthesis of proteins and nucleic acid (Adenocarcinoma et al. 2014; Length 2007; Mumtaz et al. 2002; Skipper et al. 2016; Yedjou and Tchounwou 2007). Against cadmium exposure many stress response system are activated e.g. Heat shock, oxidative stress, cold shock, stringent response, SOS etc. and it is proved by using 2-D gel-electrophoresis studies. According to in vitro studies cadmium concentrations from 0.1 to 10 μm can induce free radical dependent DNA destruction and cytotoxic effect.

Cadmium being a weak mutagen alters signal transduction induces production of inositol polyphosphate, high amounts in cytosolic free calcium level in different cell types and restricting calcium channels. At lesser concentration (1–100 μm) cadmium sticks to protein leading to 70 protein degradation, poor DNA repair, and increases the cytokines and proto-oncogenes as for c-hyc, c-jun, c-fos and gear ups expression of various gene consisting of metallothioneins glutathione transfers heat shock protein, heme oxygenase, acute-phase reactants and DNA polymerase β (Kazemipour et al. 2008; Morin et al. 2007). At a concentration of 4 mg/kg body weight, male reproduction changes as in mice model. Cadmium is considered to be a human carcinogen as it had been found that people suffering from lung cancer were exposed to cadmium also that data shows pulmonary systems, as the primary target site of exposure to cadmium. All the cancerous heavy metals are found to cause DNA damage through base pair mutation, deletion, or oxygen radical attack on DNA (Aziz et al. 2008; Kazemipour et al. 2008; Morin et al. 2007; Skipper et al. 2016).

7.4 Mercury

The molecular system of toxicity of mercury are relied upon its biological characteristics and chemical activities which refer that oxidative stress is responsible in its toxicity (Clarkson and Magos 2006). Oxidative stress of mercury has exhibited that mechanisms of sulfhydryl reactivity. Hg^{2+} and MeHg make covalent bonds with cysteine residues of protein and consume cellular anti-oxidants. Consumption of mercury compounds causes oxidative damage by gathering reactive oxygen species (ROS) which generally gets removed by cellular anti-oxidants (Clarkson and Magos 2006; Goyer et al. 2004; Jan et al. 2015; Patil et al. 2013; Singh et al. 2011). In eukaryotes, the synthesis of ROS is done in the mitochondria by normal metabolism (Clarkson and Magos 2006; Goyer et al. 2004; Jan et al. 2015; Patil et al. 2013; Singh et al. 2011; Stohs and Bagchi 1995). Inorganic mercury cause increase in synthesis of ROS through inducing glitch in oxidative phosphorylation and electron transport. Mercury causes underdeveloped shedding of electrons to molecular oxygen which results in an increase in production of ROS by increasing of electron transfer in electron transport frame (Chioma et al. 2017; Odewabi and Ekor 2017).

Organic mercury compounds are found to cause growth in intracellular calcium by advancing the influx of calcium against extracellular medium and mobilizing intracellular stores. Mercury compounds causes increased level of 3-4 methylenedioxyamphetamine (MDA) in livers, kidney, and lungs. Carcinogenesis is found to have its stages viz. initiation, promotion, progression, followed by metastasis. Exposure to mercury has been a doubtful topic (Clarkson 1993; Khoury et al. 2015; Yedjou et al. 2015). There are some studies which assure the genotoxic potential of mercury while other deny (Puccetti et al. 2000; Tchounwou et al. 2003). Mercury causes production of ROS which is known to lead to DNA damage in cells (Al-azzawie et al. 2013), a method that is known to lead to the carcinogenesis procedure. Though mercury and its compounds are not mutagenic in bacterial assays, inorganic mercury is found to cause mutational events in eukaryotic cells (Schurz et al. 2000). People consuming contaminated fish which is intoxicated by methyl mercury are found to have higher Glutathione levels. Despite of all, the studies show the chronic intake of mercury causing DNA damage also it can be cell specific as well as species specific.

7.5 Lead

Lead toxicity possess a lot of severed unfavorable impacts in both in adults and children's populations (Goyer 1993; Kaul et al. 1999). In children's it causes blood poisoning and diminished intelligence, hindered neurobehavioral development, diminished hearing sharpness, discourse and dialect handicaps, development implement, poor capacity to focus, and hostile to social and persistent practices (Alghazal et al. 2008; IARC 2006). In grown up population defects like diminished Sperm check in Men, abortions or pre-nature births in Women are caused by high lead exposure (Yedjou and Tchounwou 2007).

In acute exposure, Lead can caused damage to kidney, brain, and various gastrointestinal disease and it's chronic exposure through an adverse impact Vitamin D metabolism, blood pressure and CNS (Awasthi et al. 1996; Heipieper et al. 1996). Lead having an ability that it can mimic or inhibit the action of calcium by the way it can incorporate itself in place of calcium inside the skeleton and then interact with various biological molecules like proteins etc. and by acquiring a number of mechanics it interrupts their function. When amide and sulfhydryl groups of enzyme from a complex with lead it changes their configuration and decreased their actions or activities (Awasthi et al. 1996; Heipieper et al. 1996; Village 2005). In human externalization of phosphatidylserine and turn on of caspase-3, damage of DNA, transcriptional turn on of stress genes, oxidative stress and cell deaths are events that are associated with many cellular and molecular processes observed apoptosis and toxicity and are involved due to lead as reported by various research studies (Kazemipour et al. 2008; Patil et al. 2013; Village 2005).

8 Special Considerations

8.1 Children

Children are at much higher risk of being caught by environmental hazards. The possible reasons for this are—drinking more water, breathing more air, eating more food per unit weight also they are more in contact with the floor and they touch and put it in their mouth which seems them attractive (Hotz et al. 1999). The major difference between a children mechanism than the adults is their immune system. While adults have much developed immune system than children thus they are more prone to be caught up by the diseases. Children playing outside are often found to suffer more from air pollution (Alfvén et al. 2000).

Mercury is present in high amount in fishes of fresh water and ocean through disposal of mercury in water. Thus, consumption of fishes can damage the brain effect the memory of a person. This can be illustrated by a case in Minamata Bay, Japan in 1960s where discharge of large amount of mercury in the bay caused contamination of the fishes and ingestion of those fishes by pregnant women resulted in death of 41 infants and 30 found to be born with brain injury hereby, certifying the bad impacts of mercury on a child's health (Jarup et al. 1995). Children don't have developed blood-barrier like as in adults. So, the inorganic lead can pass through the blood-barrier in children making them exposed to the diseases caused by lead toxicity.

Cadmium has a half-life of 10–30 years in bones and kidneys thus children suffer more from cadmium toxicity from its exposure (Nishijo et al. 1995). Pregnant women who smoke causes serious threat to their infants since tobacco and tobacco smoke contain cadmium which can also cause cancer at its highest exposure (Duruibe et al. 2007). Thus children are at much higher risk of being exposed to toxic metals thus immediate prevention from them is the time's need. Along with all these environmental and parental factors poverty is also a major factor since children do not get proper nutrition, proper medication, and healthy environment thereby leading to chronic exposure to metal toxicity. Soil contains traces of many metals along with pesticides and many other toxic materials (Ayandiran et al. 2009). Therefore, its consumption is very unsafe to health. Many children develop habit of eating soil which if ignored can cause serious threats to life (Morin et al. 2007).

8.2   Challenges Ahead

Exposure to toxic metals not only causes serious illness and even deaths. Inadequate services, unawareness of people about the diseases from metal toxicity has made it much serious case which is needed to handled immediately (Appenroth 2010). Almost all the metals on their high exposure show similar symptoms. It is a big challenge in front to identify all the factors which make people especially children exposed of metal toxicity. This subject is not given its needed priority in medical and nursing schools as a result there are very less doctors who have intense knowledge to this subject. Thus, an urgent concern over this subject is the needed. The factors, the relation between metal exposure and risk of disease caused are the important matters to be understood in depth. Factors like smoking and obesity require a more deep inspection (Al-fartusie and Mohssan 2017; Jaishankar et al. 2014). One of the biggest challenges is to understand the carcinogenetic impacts of some heavy metals on their severe exposure. Monitoring and establishing the measures to control over metal toxicity is a big challenge and will require additional resources and inter sectoral collaboration.

8.3   Eco-Friendly Ways to Remove Heavy Metal Toxicity

Water is life-essential resource. Water is being used by each and every living-organism that is present on this earth because of this element (Paknikar et al. 2003; Volesky et al. 1995). Chemically it is oxygen and hydrogen but its application is very broad. Because of a poor life-style and management this resource is being polluted day by day. One of the reasons which need our strong concern is the heavy metal toxicity in water bodies. Lead, chromium, mercury, uranium, selenium, zinc, arsenic, cadmium, silver, gold and nickel are the metals considered as threat if occurs in a large quantity in the living organisms (Paul et al. 2006; Yan and Viraraghavan 2000). In the natural environment the sediment and ores are primary sites of the heavy metals where these are found in immobilized form. However, we have observed an increment in the levels of heavy metals that are depositing itself in our aquatic and terrestrial environment and the reason behind this is the several human undertakings (Wilke et al. 2006; Duruibe et al. 2007; Morin et al. 2007). For example, industrial activities and ore mining that has disturbed the natural biogeochemical circle. When these pollutants are liberated out in the absence of a regular treatment causes a trouble for natural system as well as for the public health (Ayandiran et al. 2009; Wilke et al. 2006). These heavy metals are non-biodegradable and remains as it is or constant with the passage of time. Metals can enter into the food web by a process known as leaching in which the metals are extracted away from the dumped waste materials, polluted soils and water. This process leads to another important phenomenon that is bio-magnification where these toxic metals get incorporated in food chains.

We can also use a word bio-accumulation in which certain substances or chemical gets deposited inside an organism or plants (Paul et al. 2006; Yan and Viraraghavan 2000). These Heavy metals have ability that they can bind with protein molecules and can restrict the process of DNA replication which further blocks the process of cell-division. Therefore, to prevent this health risk we need to discard these toxic metals from waste water/polluted water before its further disposal. Heavy metals should be removed from the waste water before their disposal in order to prevent health related risks. The various sources of heavy metal poisoning are urban industrial aerosols, solid wastes from animals, mining activities, industrial and agricultural chemicals. Acid rain and break down of soil and rock into water also contaminates the water (Paknikar et al. 2003; Volesky et al. 1995).

To purify the contaminated water resources, there are several technologies viz. reverse osmosis, electro dialysis, ultra-filtration, ion-exchange, chemical precipitation, phytoremediation etc. However, these methods are not subjected for total removal of metal removal (Duruibe et al. 2007). Since the present technologies have various disadvantages therefore we need some cost-effective alternatives technologies. Recently Biomass has been emanated as another waste water treatment process and it is a cost-effective and eco-friendly method. Biosorption is defined as "a non-directed phyicochemical interaction" which can occur in midst metal and microbial cells. It can be used to treat contaminated water also it has several pros over other methods like chemical/biological sludge economical, regeneration of biosorbent making it is possible to take out metal from contaminated water.

Solvent is attracted and bounded to sorbate with various mechanisms since the sorbent has a higher affinity. This process keeps on going until equilibrium is established among quantity of solid-bound sorbate species and its part left in the solution (Yan and Viraraghavan 2000).

A fine biosorbent leads to fruitful biosorption (Table 13). However starting from selection of types of biomass followed by prior treatment confinement is done so

Table 13 Shows the biosorption mechanism of copper in different regents

Mechanism	Biosorption of	Regent used
Physical adsorption	Copper	Bacterium Zoogloea ramigera and Alga cholera
Ion-exchange	Copper	Fungi Ganoderma lucidum and Aspergillus niger
Complexation	Copper	C. vulgaris, and Z. ramigera

as to gain productivity of metal uptake and hence removing the adsorbed metal, by desorption process so as the biosorbent can be reiterated for other operations (Paknikar et al. 2003).

8.4 Toxic Heavy Metals and Undeclared Drugs

Asian Herbal Medicines (AHMs) are becoming more pronounced in most the developed countries (Ko 1999). AHM's are not supplied as medicine since because of proper information about pharmacology and toxic properties are disguised (Cosyns et al. 1999; Napolitano 2001). It is a crucial matter to be looked as AHM's contain heavy metals or undeclared drugs. In India, a case revealed that out of 12 cases of poisoning in drug in taking, 9 were caused due to herbal medicines which contained inappropriate amount of heavy metals. A recent report by Indian authors brought out that 31 ayurvedic traditional medicines contained mercury, out of which 30 contained it in amount more than as set up by the standards i.e. 1 ppm. These data bring up about the real picture of herbal Ayurvedic medicines in India. Thus, it should be over looked.

8.5 Chinese Herbal Ayurvedic Medicines

In China, from time to time, various case and series of incidents of heavy metals related to use of traditional Chinese medicine have been published. In California various Chinese herbal medicines have been banned in the retail stores. However, heavy metals are not only contaminant present in herbal remedies; they are even associated with contaminants like herbicides, pesticides, micro-organisms or mycotoxins, insects or undeclared herbal constituents. In Belgium contamination of heavy metals because of plants of Aristolochia species resulted in plague of subacute intestinal nephropathy which caused kidney transplantation of many of the patients (Ernst 2002; Ko 1999; Koneman and Roberts 2002; Saper et al. 2008). In various case reports published by different countries viz. Australia, Belgium, China, Netherlands, New Zealand, UK and USA states that adulteration of TCM's with some synthetic drug causes health problems to user some of which are fatal (Barnes 2003; Cosyns et al.

1999; Ernst 2004; Keane et al. 1999; Linde et al. 2001). The symptoms to these altered herbal remedies may appear or may not.

8.6 Concern About Safety of Asian Herbal Medicines

The above data reveals the critical situation of present which could even become worse if not handled today. The herbal medicine that we take to cure the disease is itself causing diseases because of being altered by various adulterants. Thus the current need is to restrict the supply of contaminated herbal medicines (Barnes 2003; Keane et al. 1999).

8.7 Measures to Be Taken by Every Patient with Reference to Use of Herbal Medicine

- The Herbal remedies must be regarded as medicines.
- The Herbal remedies should be in taken by doctor's prescription and dosages should also be followed.
- Long term use of these medicines should be prevented.
- If some undesired symptoms are observed after ingestion to the herbal medicine. Immediately stop its use and report it to your doctor.
- Be careful of the adulterated herbal medicine.
- Buy it only from reputed stores.
- Pregnant women and young children should not in take herbal medicine.

9 Conclusion

Some of the heavy metals are directly associated with cancer initiation and progression through suppressing immune system and altering cancer signaling pathways. Heavy metals which are associated with cancer are: arsenic, uranium mercury, lead, cadmium and aluminum etc. These deadly and silent invaders cause suppression and/or deregulation of the immune system, leading to cancer initiation and progression. Heavy metals are also linked to increased free-radical activity, DNA damage, apoptosis, cell damage, cell death, ROS and NOS generation and oxidation processes that promote cancer initiation and progression. Understanding the molecular mechanisms of heavy metal toxicity in cancer initiation and progression would be helpful to find effective therapeutic intervention for the cancer specifically induced by heavy metals.

References

Adenocarcinoma B, Tchounwou CK, Yedjou CG, Farah I, Tchounwou PB (2014) NIH Public Access 601:156–160

AERB (2004) Drinking water specifications in India. Retrieved from Mumbai: www.aerb.gov.in

Al-azzawie HF, Umran A, Hyader NH (2013) Oxidative stress, antioxidant status and DNA damage in a mercury exposure workers. Br J Pharmacol Toxicol 4:80–88

Al-fartusie FS, Mohssan SN (2017) Essential trace elements and their vital roles in human body. Indian J Adv Chem Sci 5(3):127–136

Alfvén T, Elinder CG, Carlsson MD, Grubb A, Hellström L, Persson B, Järup L (2000) Low-level cadmium exposure and osteoporosis. J Bone Miner Res 15(8):1579–1586

Alghazal MA, Lenártová V, Holovská K, Sobeková A, Falis M, Legáth J (2008) Activities of antioxidant and detoxifying enzymes in rats after lead exposure. Acta Vet Brno 77(3):347–354

Appenroth K (2010) Definition of " Heavy Metals" and their role in biological systems. In: Soil heavy metals. Soil biology, vol 19. Springer, Berlin, pp 19–30

Arif N, Yadav V, Singh S, Singh S, Ahmad P, Mishra RK et al (2016) Influence of high and low levels of plant-beneficial heavy metal ions on plant growth and development. Front Environ Sci 4:69

Awasthi S, Awasthi R, Pande VK, Srivastav RC, Frumkin H (1996) Blood lead in pregnant women in the urban slums of Lucknow, India. Occup Environ Med 53(12):836–840

Ayandiran TA, Fawole O, Adewoye SO, Ogundiran M (2009) Bioconcentration of metals in the body muscle and gut of Clarias gariepinus exposed to sublethal concentrations of soap and detergent effluent. J Cell Anim Biol 3(8):113–118

Aziz HA, Adlan MN, Ariffin KS (2008) Heavy metals (Cd, Pb, Zn, Ni, Cu and Cr(III)) removal from water in Malaysia: post treatment by high quality limestone. Biores Technol 99(6):1578–1583

Bajwa BS, Kumar S, Singh S, Sahoo SK, Tripathi RM (2017) Uranium and other heavy toxic elements distribution in the drinking water samples of SW-Punjab, India. J Radiat Res Appl Sci 10(1):13–19

Barnes J (2003) Quality, efficacy and safety of complementary medicines: fashions, facts and the future. Part 1: regulation and quality. Br J Clin Pharmacol 55:226–233

Barra R, Colombo JC, Eguren G, Gamboa N, Jardim WF, Mendoza G (2006) Persistent organic pollutants (POPs) in eastern and western South American countries. In: Reviews of environmental contamination and toxicology. Springer, New York, NY, pp 1–33

Basu A, Mahata J, Gupta S, Giri AK (2001) Genetic toxicology of a paradoxical human carcinogen, arsenic: a review. Mutat Res Rev Mutat Res 488(2):171–194

Brindha K, Elango L (2013) Occurrence of uranium in groundwater of a shallow granitic aquifer and its suitability for domestic use in southern India. J Radioanal Nucl Chem 295(1):357–367

Brochin R, Leone S, Phillips D, Shepard N, Zisa D, Angerio A (2008) The cellular effect of lead poisoning and its clinical picture. Georgetown Undergraduate J Health Sci 5(2):1–8

Browning C, The T, Mason M, Wise JP (2014) Titanium dioxide nanoparticles are not cytotoxic or clastogenic in human skin cells. J Environ Anal Toxicol 4(6):1–15

Chioma O, Emmanuel A, Peter A (2017) Accumulation and toxicological risk assessment of Cd, As, Pb, Hg, and Cu from topsoils of school playgrounds at Obio-Akpor LGA Rivers State Nigeria. Int J Sci World 5(1):38

Clarkson TW (1993) Molecular and ionic mimicry of toxic metals. Annu Rev Pharmacol Toxicol 33(1):545–571

Clarkson TW, Magos L (2006) The toxicology of mercury and its chemical compounds. Crit Rev Toxicol 36(8):609–662

Cosyns JP, Jadoul M, Squifflet JP, Wese FX, De Strihou CVY (1999) Urothelial lesions in Chinese-herb nephropathy. Am J Kidney Dis 33(6):1011–1017

Dayan AD, Paine AJ (2001) Mechanisms of chromium toxicity, carcinogenicity and allergenicity: review of the literature from 1985 to 2000. Hum Exp Toxicol 20(9):439–451

Dong Z (2002) The molecular mechanisms of arsenic-induced cell transformation and apoptosis. Environ Health Perspect 110(SUPPL. 5):757–759

Duffus JH (2002) "Heavy Metals"—a meaningless term? (IUPAC technical report). Pure Appl Chem 74(5):793–807 [National Representatives: Z. Bardodej (Czech Republic) J. Park (Korea) F. J. R. Paumgartten (Brazil)]

Duruibe JO, Ogwuegbu MO, Egwurugwu JN (2007) Heavy metal pollution and human biotoxic effects. Int J Phys Sci 2(5):112–118

Efstathiou M, Aristarchou T, Kiliari T, Demetriou A, Pashalidis I (2014) Seasonal variation, chemical behavior and kinetics of uranium in an unconfined groundwater system. J Radioanal Nucl Chem 299(1):171–175

Ernst E (2002) Adulteration of Chinese herbal medicines with synthetic drugs: a systematic review. J Intern Med 252(2):107–113

Ernst E (2004) Prescribing herbal medications appropriately. J Fam Pract 53(12):985–988

Florea AM, Büsselberg D, Carpenter D (2012) Metals and disease. J Toxicol 2012:2012–2014

Gokhale BSL (2008) Groundwater Radon-222 concentrations in Antelope Creek, Idaho: measurement and interpolation. Open Environ Bio Monit J 3:12–20

Gómez P, Garralón A, Buil B, Turrero MJ, Sánchez L, de la Cruz B (2006) Modeling of geochemical processes related to uranium mobilization in the groundwater of a uranium mine. Sci Total Environ 366(1):295–309

Goulart M, Batoréu MC, Rodrigues AS, Laires A, Rueff J (2005) Lipoperoxidation products and thiol antioxidants in chromium exposed workers. Mutagenesis 20(5):311–315

Goyer RA (1993) Lead toxicity: current concerns. Environ Health Perspect 100:177–187

Goyer R, Golub M, Choudhury H, Hughes M, Kenyon E, Stifelman M (2004) Issue paper on the human health effects of metals. In: US environmental protection agency risk assessment forum, vol 1200

Grimsrud TK, Peto J (2006) Persisting risk of nickel related lung cancer and nasal cancer among clydach refiners. Occup Environ Med 63(5):365–366

Griswold W, Martin S (2009) Human health effects of heavy metals. Environ Sci Technol 15:1–6

Hartmann HM, Monette FA, Avci HI (2000) Overview of toxicity data and risk assessment methods for evaluating the chemical effects of depleted uranium compounds. Hum Ecol Risk Assess 6(5):851–874

Heipieper HJ, Meulenbeld G, Van Oirschot Q, De Bont JAM (1996) Effect of environmental factors on the trans/cis ratio of unsaturated fatty acids in pseudomonas putida S12. Appl Environ Microbiol 62(8):2773–2777

Hodson ME (2004) Heavy metals—geochemical bogey men? Environ Pollut 129(3):341–343

Hong YS, Song KH, Chung JY (2014) Health effects of chronic arsenic exposure. J Prev Med Public Health 47(5):245–252

Hotz P, Buchet JP, Bernard A, Lison D, Lauwerys R (1999) Renal effects of low-level environmental cadmium exposure: 5-year follow-up of a subcohort from the cadmibel study. Lancet 354:1508–1513

Hu Z, Gao S (2008) Upper crustal abundances of trace elements: a revision and update. Chem Geol 253(3):205–221

IAEA (2015) Protection of the public against exposure indoors due to radon and other natural sources of radiation. In: IAEA safety standards for protecting people and the environment, vol SSG-32. IAEA, Vienna, p 112

IARC (2006) Inorganic and organic lead. IARC monographs on the evaluation of carcinogenic risks to humans, 87

Ilyin I, Berg T, Dutchak S, Pacyna J (2004) Heavy metals. EMEP assessment part I European perspective. Norwegian Meteorological Institute, Oslo, Norway, pp 107–128

Jaishankar M, Tseten T, Anbalagan N, Mathew BB, Beeregowda KN (2014) Toxicity, mechanism and health effects of some heavy metals. Interdisc Toxicol 7(2):60–72

Jan AT, Azam M, Siddiqui K, Ali A, Choi I, Haq QMR (2015) Heavy metals and human health: mechanistic insight into toxicity and counter defense system of antioxidants. Int J Mol Sci 16(12):29592–29630

Järup L (2003) Hazards of heavy metal contamination. Br Med Bull 68:167–182

Jarup L, Persson B, Elinder CG (1995) Decreased glomerular filtration rate in solderers exposed to cadmium. Occup Environ Med 52(12):818–822

Jhaa N, Noditi M, Nilsson R, Natarajana T (1992) Genotoxic effects of sodium arsenite on human cells. Mutat Res 284(2):215–221

Kaul B, Sandhu RS, Depratt C, Reyes F (1999) Follow-up screening of lead-poisoned children near an auto battery recycling plant, Haina, Dominican Republic. Environ Health Perspect 107(11):917–920

Kawada T (2016) Predictive validity of a specific questionnaire for psychiatric morbidity and suicidal ideation. J Formos Med Assoc 115(11):1019–1020

Kazemipour M, Ansari M, Tajrobehkar S, Majdzadeh M, Kermani HR (2008) Removal of lead, cadmium, zinc, and copper from industrial wastewater by carbon developed from walnut, hazelnut, almond, pistachio shell, and apricot stone. J Hazard Mater 150(2):322–327

Keane FM, Munn SE, du Vivier AW, Taylor NF, Higgins EM (1999) Analysis of Chinese herbal creams prescribed for dermatological conditions. BMJ (Clin Res Ed) 318(7183):563–564

Khatri P, Sirota M, Butte AJ (2012) Ten years of pathway analysis: current approaches and outstanding challenges. PLoS Comput Biol 8(2):e1002375

Khoury EDT, Da Silva Souza G, Da Costa CA, De Araújo AAK, De Oliveira CSB, De Lima Silveira LC, Da Conceição NPM (2015) Somatosensory psychophysical losses in inhabitants of riverside communities of the Tapajós River Basin, Amazon, Brazil: exposure to methylmercury is possibly involved. PLoS ONE 10(12):1–19

Kim HS, Kim YJ, Seo YR (2015) An overview of carcinogenic heavy metal: molecular toxicity mechanism and prevention. J Cancer Prev 20(4):232–240

Kippler M, Tofail F, Gardner R, Rahman A, Hamadani J, Bottai M, Vahter M (2012) Maternal cadmium exposure during pregnancy and size at birth: a prospective cohort study. Environ Health Perspect 120(2):284–289

Ko R (1999) Adverse reactions to watch for in patients using herbal remedies. West J Med (September):181–186

Koneman EW, Roberts GD (2002) Your lab focus. Lab Med 33(12):437–445

Konietzka R (2015) Gastrointestinal absorption of uranium compounds—a review. Regul Toxicol Pharmacol 71(1):125–133

Kumar A, Usha N, Mishra MK, Tripathi RM, Rout S, Jaspal S (2011) Risk assessment for natural uranium in subsurface water of Punjab State, India. Hum Ecol Risk Assess 17:381–393

Lee CP, Lee YH, Lian IB, Su CC (2016) Increased prevalence of esophageal cancer in areas with high levels of nickel in farm soils. J Cancer 7(12):1724–1730

Length F (2007) Heavy metal pollution and human biotoxic effects. Int J Phys Sci 2(5):112–118

Liesch T, Hinrichsen S, Goldscheider N (2015) Uranium in groundwater—fertilizers versus geogenic sources. Sci Total Environ 536:981–995

Linde K, Vickers A, Hondras M, Ter Riet G, Thormählen J, Berman B, Melchart D (2001) Systematic reviews of complementary therapies—an annotated bibliography. Part 1: acupuncture. BMC Complement Altern Med 1:3

Liu Y, Guyton KZ, Gorospe M, Xu Q, Lee JC, Holbrook NJ (1996) Differential activation of ERK, JNK/SAPK and P38/CSBP/RK map kinase family members during the cellular response to arsenite. Free Radic Biol Med 21(6):771–781

Lobo V, Patil A, Phatak A, Chandra N (2010) Free radicals, antioxidants and functional foods: impact on human health. Pharmacognosy Rev 4(8):118

Mamtani R, Stern P, Dawood I, Cheema S (2011) Metals and disease: a global primary health care perspective. J Toxicol 2011(319136):1–11

Mazariegos M, Hambidge KM, Westcott JE, Solomons NW, Raboy V, Das A, Krebs NF (2010) Neither a zinc supplement nor phytate-reduced maize nor their combination enhance growth of 6- to 12-month-old guatemalan infants. J Nutr 1–4(9):1041–1048

Morin S, Vivas-Nogues M, Duong TT, Boudou A, Coste M, Delmas F (2007) Dynamics of benthic diatom colonization in a cadmium/zinc-polluted river (Riou-Mort, France). Fundam Appl Limnol/Archiv Für Hydrobiol 168(2):179–187

Mumtaz MM, Tully DB, El-Masri HA, De Rosa CT (2002) Gene induction studies and toxicity of chemical mixtures. Environ Health Perspect 110(SUPPL.6):947–956

Napolitano V (2001) Complementary medicine use by Mexican migrants in the San Francisco Bay Area. West J Med 174(3):203–206

Nishijo M, Nakagawa H, Morikawa Y, Tabata M, Senma M, Miura K, Nogawa K (1995) Mortality of inhabitants in an area polluted by cadmium: 15 year follow up. Occup Environ Med 52(3):181–184

O'Brien TJ, Ceryak S, Patierno SR (2003) Complexities of chromium carcinogenesis: role of cellular response, repair and recovery mechanisms. Mutat Res Fundam Mol Mech Mutagenesis 533(1–2):3–36

Odewabi AO, Ekor M (2017) Levels of heavy and essential trace metals and their correlation with antioxidant and health status in individuals occupationally exposed to municipal solid wastes. Toxicol Ind Health 33(5):431–442

OEHHA (Office of Environmental Health Hazard Assessment Agency California Environmental Protection) (2001) Public health goals for chemicals in drinking water: NICKEL (August)

Paknikar KM, Pethkar AV, Puranik PR (2003) Bioremediation of metalliferous wastes and products using inactivated microbial biomass. Indian J Biotechnol 2(3):426–443

Patil YP, Pawar SH, Jadhav S, Kadu JS (2013) Biochemistry of metal absorption in human body: reference to check impact of nano particles on human being. Int J Sci Res Publ 3(4):1–5

Paul S, Bera D, Chattopadhyay P, Ray L (2006) Biosorption of Pb (II) by Bacillus cereus M1 16 immobilized in calcium alginate gel. J Hazard Subst Res 5(1):2

Prüss-Ustün A, Vickers C, Haefliger P, Bertollini R (2011) Knowns and unknowns on burden of disease due to chemicals: a systematic review. Environ Health 10(1):9

Puccetti E, Güller S, Orleth A, Brüggenolte N, Hoelzer D, Ottmann OG, Ruthardt M (2000) BCR-ABL mediates arsenic trioxide-induced apoptosis independently of its aberrant kinase activity. Can Res 60(13):3409–3413

Radespiel-Tröger M, Meyer M (2013) Association between drinking water uranium content and cancer risk in Bavaria, Germany. Int Arch Occup Environ Health 86(7):767–776

Rooney JPK (2007) The role of thiols, dithiols, nutritional factors and interacting ligands in the toxicology of mercury. Toxicology 234(3):145–156

Samet JM (2011) Radiation and cancer risk: a continuing challenge for epidemiologists. Environ Health 10(1):541–549

Sankhla MS, Kumari M, Nandan M, Kumar R, Agrawal P (2016) Heavy metals contamination in water and their hazardous effect on human health: a review. Int J Curr Microbiol Appl Sci 5:759–766

Saper RB, Phillips RS, Sehgal A, Khouri N, Davis RB, Paquin J, Kales SN (2008) Lead, mercury, and arsenic in US- and Indian-manufactured ayurvedic medicines sold via the internet. JAMA J Am Med Assoc 300(8):915–923

Schurz F, Sabater-Vilar M, Fink-Gremmels J (2000) Mutagenicity of mercury chloride and mechanisms of cellular defence: the role of metal-binding proteins. Mutagenesis 15:525–530

Singh NP, Burleigh DP, Ruth HM, Wrenn ME (1990) Daily U intake in Utah residents from food and drinking water. Health Phys 59(3):333–337

Singh R, Gautam N, Mishra A, Gupta R (2011) Heavy metals and living systems: an overview. Indian J Pharmacol 43(3):246

Skipper A, Sims JN, Yedjou CG, Tchounwou PB (2016) Cadmium chloride induces DNA damage and apoptosis of human liver carcinoma cells via oxidative stress. Int J Environ Res Public Health 13(1):1–10

Soignet BSL, Frankel SR, Douer D, Tallman MS, Kantarjian H, Calleja E, Warrell RP (2001) United States multicenter study of arsenic trioxide in relapsed acute promyelocytic leukemia. J Clin Oncol Official J Am Soc Clin Oncol 19(18):3852–3860

Stevens JJ, Graham B, Walker AM, Tchounwou PB, Rogers C (2010) The effects of arsenic trioxide on DNA synthesis and genotoxicity in human colon cancer cells. Int J Environ Res Public Health 7(5):2018–2032

Stohs SJ, Bagchi D (1995) Oxidative mechanisms in the toxicity of metal ions. Free Radic Biol Med 18(2):321–336

Su C, Yang H, Huang S, Lian I (2007) Distinctive features of oral cancer in Changhua county: high incidence, buccal mucosa preponderance, and a close relation to betel quid chewing habit. J Formos Med Assoc 106(3):225–233

Takahashi M, Barrett JC, Tsutsui T (2002) Transformation by inorganic arsenic compounds of normal Syrian hamster embryo cells into a neoplastic state in which they become anchorage-independent and cause tumors in newborn hamsters. Int J Cancer 99(5):629–634

Tchounwou PB, Patlolla AK, Centeno JA (2003) Invited reviews: carcinogenic and systemic health effects associated with arsenic exposure—a critical review. Toxicol Pathol 31(6):575–588

Tchounwou PB, Centeno JA, Patlolla AK (2004) Arsenic toxicity, mutagenesis, and carcinogenesis—a health risk assessment and management approach. Mol Cell Biochem 255(1–2):47–55

Tchounwou PB, Yedjou CG, Patlolla AK, Sutton DJ (2012a) Molecular, clinical and environmental toxicology, vol 101, Springer, Basel, pp 1–30

Tchounwou PB, Yedjou CG, Patlolla AK, Sutton DJ (2012b) Heavy metal toxicity and the environment. In: Molecular, clinical and environmental toxicology. Springer, Basel, pp 133–164

Velma V, Tchounwou PB (2011) NIH Public Access 698:43–51

Verin AD, Cooke C, Herenyiova M, Patterson CE, Garcia JG (1998) Role of Ca2+/calmodulin-dependent phosphatase 2B in thrombin-induced endothelial cell contractile responses. Am J Physiol 275(4 Pt 1):L788–L799

Village G (2005) Lead exposure in children: prevention, detection, and management. Pediatrics 116(4):1036–1046

Volesky B, Holan ZR, About M, Article T (1995) Biosorption of heavy metals. Biotechnol Prog 11(3):235–250

Walker AM, Stevens JJ, Ndebele K, Tchounwou PB (2010) Arsenic trioxide modulates DNA synthesis and apoptosis in lung carcinoma cells. Int J Environ Res Public Health 7(5):1996–2007

Wen CP, Tsai SP, Cheng TY, Chen CJ, Levy DT, Yang HJ, Eriksen MP (2005) Uncovering the relation between betel quid chewing and cigarette smoking in Taiwan. Tobacco Control 14(SUPPL. 1):16–22

Wilke A, Buchholz R, Bunke G (2006) Selective biosorption of heavy metals by algae. Environ Biotechnol 2:47–56

Yamashoji S, Isshiki K (2001) Rapid detection of cytotoxicity of food additives and contaminants by a novel cytotoxicity test, menadione-catalyzed H_2O_2 production assay. Cytotechnology 37(3):171–178

Yan G, Viraraghavan T (2000) Effect of pretreatment on the bioadsorption of heavy metals on Mucor rouxii. WATER SA-PRETORIA 26(1):119–124

Yedjou CG, Tchounwou PB (2007) N-acetyl-l-cysteine affords protection against lead-induced cytotoxicity and oxidative stress in human liver carcinoma (HepG2) cells. Int J Environ Res Public Health 4(2):132–137

Yedjou CG, Milner JN, Howard CB, Tchounwou PB (2010) Basic apoptotic mechanisms of lead toxicity in human leukemia (Hl-60) cells. Int J Environ Res Public Health 7(5):2008–2017

Yedjou CG, Tchounwou HM, Tchounwou PB (2015) DNA damage, cell cycle arrest, and apoptosis induction caused by lead in human leukemia cells. Int J Environ Res Public Health 13(1):56

Yuan T, Lian I, Tsai K, Chang T, Chiang C, Su C, Hwang Y (2011) Science of the total environment possible association between nickel and chromium and oral cancer: a case—control study in central Taiwan. Sci Total Environ 409(6):1046–1052

Zhang YL, Zhao YC, Wang JX, Zhu HD, Liu QF, Fan YG, Fan TQ (2004) Effect of environmental exposure to cadmium on pregnancy outcome and fetal growth: a study on healthy pregnant women in China. J Environ Sci Health Part A 39(9):2507–2515

Zhang R, Ma A, Urbanski SJ, McCafferty DM (2007) Induction of inducible nitric oxide synthase: a protective mechanism in colitis-induced adenocarcinoma. Carcinogenesis 28(5):1122–1130

Zofkova I, Davis M, Blahos J (2017) Trace elements have beneficial, as well as detrimental effects on bone homeostasis. Physiol Res 66:391–402

Burden of Occupational and Environmental Hazards of Cancer

Meenu Gupta and Anupam Dhasmana

Abstract Ecological studies showed the association of exposure to carcinogens present surroundings an indoor and outdoor environment. The International Agency for Research on Cancer (IARC) has classified arsenic, asbestos, benzene, radon gas etc. into group 1 carcinogens. In many countries, pollution is rising due to trend of increasing industrialization and urbanization, occupational exposure to asbestos and chemical carcinogens. Mostly in developing countries, women are traditionally leader in cooking but due to frequent use of biomasses and wood fuel in poor ventilated houses, they are exposed to indoor air pollutants. However, not only in developing countries, but also in developed countries like United States facing serious smoking problem where after smoking, radon is the second common cause of bronchogenic cancer. Smoking and radon exposure has synergistic effect on carcinogenesis. Adequate legislation like banning, elimination or substitutions of carcinogens in industries along with the public education can help in reduction of burden of the environmental and occupational cancer. This chapter is in process to explore the exiting occupational and environmental hazards present in the environment and causing several health diseases.

Keywords Environment · Carcinogens · Asbestos · Radon

1 Introduction

Environment is everything outside the body that interacts with humans or living matter. Environment is derived from the French word "Environ" which means, "surrounding". For human health and well-being, a clean environment is necessary. Inter-

M. Gupta (✉)
Department of Radiation Oncology, Himalayan Institute of Medical Sciences and Cancer Research Institute, Swami Rama Himalayan University, Dehradun, India
e-mail: meenugupta.786@rediffmail.com

A. Dhasmana
Department of Biosciences, Himalayan Institute of Medical Sciences and Cancer Research Institute, Swami Rama Himalayan University, Dehradun, India

© Springer Nature Switzerland AG 2019
K. K. Kesari (ed.), *Networking of Mutagens in Environmental Toxicology*, Environmental Science,
https://doi.org/10.1007/978-3-319-96511-6_4

actions between environment and human health involves various pathways. Some substances found in environment can potentially lead to cancer, which has been defined as an uncontrolled division of cells. There are other complex factors in addition to carcinogens in the environment, which may lead to the development of cancer, including lifestyle and genetic makeup. Cancer is the second leading cause of death globally. Globocan report showed that cancer is responsible for an estimated 9.6 million deaths in 2018 (Bray et al. 2018). WHO reported that 19% of all cancer cases are attributable to the environment, including the workplace (IEPH 2016). The burden of environment related cancer deaths is 1.7 million deaths annually (Bray et al. 2018).

International Agency for Research on Cancer (IARC) classifies various compounds or physical factors, which can cause cancer (Cogliano et al. 2011).

Group 1: "Carcinogenic to humans"—There is enough evidence to conclude that it can cause cancer in humans. Examples includes asbestos, benzene and ionizing radiation.
Group 2A: "Probably carcinogenic to humans"—There is strong evidence of causing cancer in humans, but at present, it is not conclusive. Examples includes diesel engine exhaust, formaldehyde and PCBs.
Group 2B: "Possibly carcinogenic to humans"—There is some evidence causing cancer in humans but at present, it is far from conclusive. Examples includes styrene and gasoline exhaust.
Group 3: "Unclassifiable as to carcinogenicity in humans"—There is no evidence at present causing cancer in humans. Examples include anthracene, caffeine and fluorescent lighting.
Group 4: "Probably not carcinogenic to humans"—There is strong evidence that not causing cancer in humans. Example is Caprolactam.

In the process of cancer formation, there is an activation of proto-oncogenes and the inactivation of tumor suppressor genes. Studies showed that combined effect of genetic and external factors acting concurrently and sequentially results in cancer formation (Lodish et al. 2000).

2   Cancer Risk from Environmental Exposure to Arsenic

Arsenic, a naturally occurring metalloid, is present in rock, soil, air, water and is constituted in animals and plant kingdom. The source of consumption of arsenic inside our body enters through the breathing air, drinking water, and the consumption of food. Inorganic compounds of arsenic found in industry, building products and arsenic-contaminated water has been linked to cancer (Arsenic cycle is shown in Fig. 1). Main source of environmental exposures to arsenic in some regions of the world is drinking water. An increased levels of arsenic occurs naturally in drinking water in various parts of Taiwan, Bangladesh, Japan and Western South America. Ground source of water like wells are rich source of arsenic as compared to water from lakes or reservoirs (Surface water). Based on the evidences from human

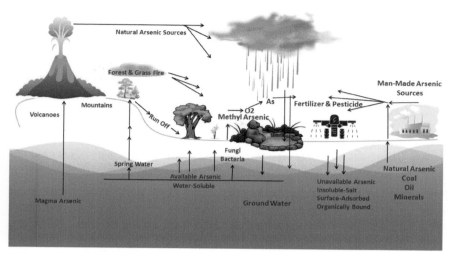

Fig. 1 The arsenic cycle in nature

studies, expert agencies like the "International Agency for Research on Cancer" (IARC) classifies arsenic as well as inorganic arsenic compounds as "carcinogenic to humans". Arsenic exposure can result in carcinoma of the lung, urinary bladder and skin. Chronic exposure to this element usually results in carcinogenesis with a latency period of 30–50 years (Cogliano et al. 2011; Ratnaike 2003). Data published by Oberoi et al. (2014), showed that in every single year 9129–119,176 additional cases of urinary bladder cancer, 11,844–121,442 cases of bronchogenic cancer, and 10,729–110,015 cases of skin cancer globally are attributable to inorganic consumption of arsenic in food products.

The maximum contamination level of arsenic allowed in drinking water in United Kingdom and United States of America is 10 μg/L (Andrew et al. 2017). The WHO confirmed a guideline level of 10 μg/L for inorganic arsenic in drinking water. Elevated arsenic concentrations in groundwater is found in certain locations like in Bangladesh. Countries like Bangladesh have adopted a guideline of 50 μg/L due to non availability of alternative water (Smith and Steinmaus 2009). In rural areas of Bangladesh, people depends on untreated groundwater for their drinking water consumption. Arsenic contaminated water when used for irrigation in agricultural settings also has an impact on crop yields. Rice is the primary source of calories in this country. A survey was done for estimation of arsenic levels in paddy soils of Bangladesh and results showed that in those zones where groundwater used for irrigation has high arsenic concentrations, and also where these tube wells were functional for the prolonged period of time, arsenic levels were elevated in these zones (Meharg and Rahman 2003). Study of nine hundred and one Polished *White Rice Grain* samples sourced from ten countries and four continents estimated cancer risks by multiplying daily arsenic intakes by the slope of internal cancer risk proposed by the Environmental Protection Agency. For a fixed consumption of "100" g intake of rice daily, median excess internal cancer risks were 7 in 10,000 for India, 15 in 10,000 for China, and 22 in 10,000 for Bangladesh (Mukherjee et al. 2006). In parts of Eastern India, arsenicosis due to environmental exposure is a serious public health concern. A case of double malignancy of bronchogenic as well as skin squamous cell carcinoma in 49 years old male has been reported by Sinha et al. (2016). Source of drinking water for this patient was from tube well which contains an increased level of arsenic (0.125 mg/L), where arsenic test was found positive in his hair and nail samples. He received chemotherapy (gemcitabine/carboplatin), followed by radiotherapy. There was complete response seen on radiological images. This case is a warning sign for public health concern for arsenic exposure in this part of India.

Prevention and Control: Provision of safe water supply for drinking, cooking and irrigation of crops can help the affected communities. Water with low arsenic can be utilized for food preparation, drinking and cultivation, and for bathing and washing clothes, high arsenic water can be used. Low-arsenic water with higher-arsenic water can be blended together for an acceptable arsenic concentration level. Rain water storage can be safe. Painting of tube wells or hand pumps with different colours symbolizing high or low arsenic concentration can help the people. Various techniques like oxidation, coagulation-precipitation, absorption and ion exchange are used for arsenic removal systems. The highly recommendable preventive and control methods can be considered by providing awareness through education to the communities living around affected areas (Flanagan et al. 2012; Diaz and Arcos 2015).

3   Asbestos (IARC Group 1 Carcinogen)

Asbestos is bundles of six naturally occurring longer thinner fibres of the mineral silicate of "serpentine" and "amphibole" series. These are serpentine mineral chrysotile (White asbestos) and 5 amphibole minerals which are actinolite, amosite (Brown asbestos), anthophyllite, crocidolite (Blue asbestos) and tremolite. All the commercial form of Asbestos are declared as human carcinogens by IARC. Asbestos market is captured and maintained by the urban industries involved in mining, manufacturing and handling of asbestos containing products. Epidemiological studies showed that high incidence of lung cancer, pleural, peritoneal mesotheliomas and gastrointestinal tract cancers were reported in the group of occupationally exposed to asbestos fibers (Kim et al. 2013). Government regulations have imposed permissible limits of 2 fibers/cm^3 asbestos. Italy was the leader of asbestos production and consumer in Europe of 20th century until it was banned in 1992. Occupational exposure to asbestos occurs in mining and marketing of asbestos, shipyards, cement production, asbestos textured ceiling (Fig. 2) and textile industries (Marsili et al. 2017). India is one of eight countries that accounted for 80% of the world's asbestos consumption including Russia, China, Kazakhstan, Ukraine, Thailand, Brazil, and Iran. According

(a)   **(b)**

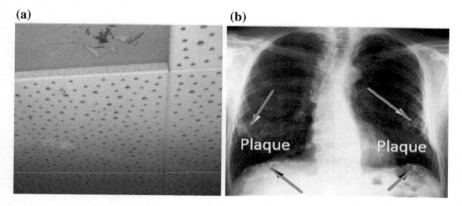

Fig. 2 **a** Asbestos texured ceiling. **b** Pleural plaques due to asbestos exposure

to US geological survey report, it was found that asbestos consumption increased in Asian countries like India, China, Kazakhstan and Ukraine.

Various studies showed the supra-additive effects of increase in lung cancer risk with asbestos, smoking and asbestosis. Lung cancer mortality among non-smokers increases with exposure to asbestos as shown by Markowitz et al. (2013). Public health challenge arises due to the decrease in asbestos usage by more developed countries and the burden is taken by less-developed countries that are continuing to use asbestos.

Prevention: Workers 'Medical Screening Examination' is must to identify early detection of risk factors and lung changes where intervention can have maximum benefit. Preplacement and periodic examination including blood, urine, exhaled air, pulmonary function tests, radiological examination of lungs to be done. An occupational exposure history and a respiratory health questionnaire must be recorded. Use of respiratory masks and protective clothing during occupational exposure can help the occupational workers. Employers must operate a medical surveillance program for all workers (Szeinu et al. 2000). Workers should be advised to quit smoking as already mentioned that smoking with simultaneous asbestos exposure has synergistic effect on likelihood of lung cancer regardless of the use of respiratory protection equipment (Hashim and Boffetta 2014). Three Es: Enforcement, Engineering and Education in the workplace can reduce the hazards.

4 Benzene Exposure

Report of "International Agency for Research on Cancer (IARC)" showed that the benzene is carcinogen substance in both animal and humans studies (McMichael 1988). Benzene is a light yellow liquid, flammable substance having aromatic odour. Route of entry into the body is through inhalation, ingestion and skin exposure.

Fig. 3 Household and commercial chemical carcinogens those accumulated indoor and outdoor environment

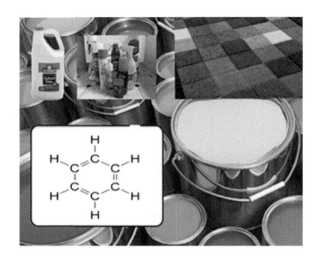

Metabolites of benzene from the liver like Benzoquinone and Mucoaldehyde causes toxicity in bone marrow. Benzene is used in organic solvent, inks in the printing industry, lubricants, dyes, cleaners, production of rubber, and pesticides in the chemical and pharmaceutical industries (Fig. 3). The benzene affect Hematopoietic Stem Cells (HSC) and differentiation steps of progenitor cells thus cause different type of haematological cancers derived from stem cell of hematopoietic system (Kauts et al. 2016). Benzene exposure occurs by breathing in air containing benzene, breathing second-hand smoke, burning coal and oil emissions, motor vehicle exhaust, and gasoline service stations evaporation. Data showed that about 50% of the exposure to benzene in the US results from tobacco smoking or from exposure to tobacco smoke. Benzene from gasoline can be permeated into the skin (Korte et al. 2000).

Prevention: Reduction of benzene exposures like vehicle exhausts emissions can be minimized by reviewing and improving the designs and continuous monitoring of engine settings. Location and design of petrol filling stations policies should be strong. Domestic use of benzene-containing products should be avoided and infants and children to be isolated from vehicle emissions in indoor settings (Nazaroff 2013). Replacement or utilization of alternative solvents in industrial processes can help in risk reduction. One example of occupational aplastic anaemia reduction occurs in a Chinese shoemaking factory in China. Here four cases of aplastic Anaemia were detected among 211 workers over an eight-month period. No further case occurs once benzene was replaced by new solvent (Issaragrisil et al. 2006). Public awareness and educational activities required in industry and domestic sites.

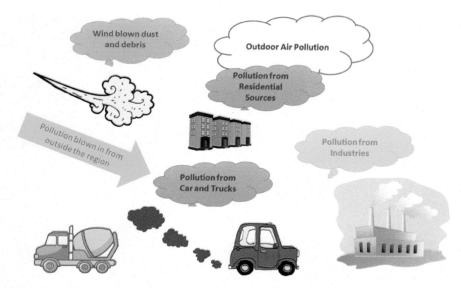

Fig. 4 Carcinogens, like diesel engine exhaust, solvents, metals, and dust are the main source of outdoor air pollution

5 Outdoor Air Pollution (OAP) in Urban Settings

Air pollution causes cancer confirmed by WHO. Reports of IARC concluded that outdoor air pollution causes cancer of lung and an increased risk for bladder cancer is linked with exposure to outdoor air pollution. Industrial sources, power plants, motor vehicles etc. release carcinogens into outdoor air (Fig. 4). In densely populated industrial urban territories, outdoor air contains human carcinogens like benzene, benzopyrene, and benzene soluble organics. The excess lung cancer risk associated with ambient air pollution is small as compared to cigarette smoking (Srogi 2007). More polluted cities like in China, OAP may contribute to as much as 10% of bronchogenic carcinoma overall, and may be a larger proportion is contributed in women who never smokes. As there is a lack of studies with robust data in developing world, most of conclusions are based on extrapolating the relative risk estimates from the ACS study to China, India, and other settings. Developing world has differences and variations in health status and composition of air pollution as compared to developed world, which results into uncertainties (Madaniyazi et al. 2015).

6 Indoor Air Pollution

Indoor living habits, such as passive smoking, cooking on solid fuel like coal, biomass, wood, crop residues and inadequate ventilation systems produce numerous

Fig. 5 Indoor air pollution by incomplete combustion of wooden/bio fuel inside ill ventilated houses

indoor air pollutants (Fig. 5). Tobacco smoke contains over 4000 chemicals in the form of particles and gases. Of the 4000 chemicals, 60 are carcinogens, including benzene, cadmium, formaldehyde and toluene. Smoking around the children results in second hand smoke (SHS). Children's are higher risk by this SHS due to respiratory rate, which is slightly more than the adult counterparts and large lung surface area. An increased amount of nicotine from second hand smoke is found deposited on household surfaces, furniture, air and dust in the homes (Burton 2011).

An hour a day in a room with a smoker is nearly a hundred times more likely to cause lung cancer in a non-smoker than 20 years spent in a building containing asbestos-Sir Richard Doll.

Globally approx. 2.8 billion people depends upon solid fuels like coal, biomass and simple stoves. In developing countries, biomass fuels in open fires from wood and cake of animal dung and traditional stoves are used for extensive periods inside poorly ventilated dwellings. Thus leading to increased levels of household air pollution (HAP) exposure. Health damaging pollutants like carbon monoxide, sulfur oxides, nitrogen oxides, aldehydes, benzene, and particulate matters and polyaromatic compounds are emitted by smoke of coal and biomasses. Indoor coal combustion emissions are human carcinogens and defined in IARC Group 1. Concentrations of polycyclic aromatic hydrocarbons and other carcinogenic compounds in the wood smoke labelled as mutagenic. Evidence of genotoxic effects are seen in subjects exposed to wood smoke (Bruce et al. 2010).

Prevention: Educational public awareness is best method of prevention. Moreover, IAP associated with solid fuel use due to poor socioeconomic conditions can be alleviated by poverty reduction. IAP is a major concern for health of women and young children, who may spend many hours close to the fire. Stoves with flues that vent smoke to the exterior, cleaner fuels (LPG or kerosene) etc. should be provided by Government to poor communities with utilization of NGO services. The Indian National Programme of improved cook stoves was established in 1983 with targets like conserving fuel, reducing smoke emissions in the cooking area, improving health conditions and improving employment opportunities for the people living in rural areas with poor socioeconomic circumstances (Jeuland and Pattanayak 2012). Along with distribution of stoves, education is required regarding operating and maintaining of these durable stoves.

7 Radon Gas

Radon is chemically inert radioactive gas that is colourless and odourless. When atoms of uranium 238 decay, due to radioactive disintegration reactions they transformed into several series of other radioactive elements. The "fifth generation" is radium, and its decay generates radon. Uranium traces are easily found in earth's rock and soil in most of the areas of United States. Depending on the underlying geology, its concentrations vary from place to place. The Environmental Protection Agency (EPA) report showed that radon is the 2nd leading cause of lung cancer in the United States which may killing 21,100 people per annum (Cao et al. 2017). The combined health effects of radon and tobacco exposure are synergistic rather than additive, so reducing either of the exposures substantially reduces lung cancer risks (Keith et al. 2012). Up to 20% of lung cancer mortality in the United States occur each year in an individuals, who have never smoked, and which may translates to about 30,000 Americans in 2017. Protracted exposure to radon is considered the most common cause of lung cancer in this population (Ou et al. 2018).

Half-life of radon gas is 3.8 days, as this is produced from rocks and soils, it has tendency to concentrate in enclosed spaces like underground mines or houses (Lugg and Probert 1997). Buildings design and construction and the quantity of radon in the underlying soil has impact on the magnitude of indoor radon concentration. Radon gas from the ground or soil into the houses can be entered due to vacuum created because of pressure differences between the house and the soil. Air pressure inside a home is often lower than the pressure in the soil especially near the basement floor slab and these air pressure differences allows the houses act as vacuum and there is the easy entry of radon gas inside the houses. Air Cracks in concrete floors and walls, construction joints, around pipe penetrations, or pores in hollow-block walls are route of entry of radon gas inside homes (Fig. 6). Radon gas released by well water during showering and other household activities is another source (Nielson et al. 1997).

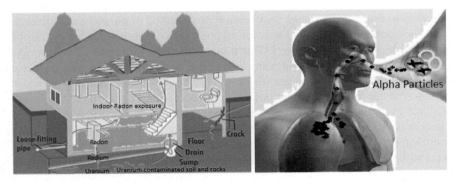

Fig. 6 Uranium contaminated rocks, soil are the source of indoor radon gas exposure, which emit alpha particles, and these particles are inhaled and absorbed by lung tissue

7.1 How Radon Causes Cancer?

Radioactive decay of radon results in emission of Alpha particles, which are ionizing radiations. It also emits some radioactive short-lived decay products. Alpha particles can penetrate the sensitive and unshielded lung cells (Fig. 6) and thus causes damage to double helical strands of DNA. Picocurie (pCi) is rate of radioactive decay of radon. Four picocuries per liter of air (4 pCi/L) is the U.S. Environmental Protection Agency (EPA) recommended action level. The EPA recommends this level when owners of houses take action for radon reduction. EPA estimates that nearly one out of every fifteen homes in the United States has radon levels above the action level. One in three houses in Minnesota have been found radon levels above the action level (Levy et al. 2015a, b). Eight Universities and a few research Institutions in India participated in co-ordinated research project sponsored by the Board of Research in Nuclear Sciences (BRNS) of Department of Atomic Energy. The results show that the radon gas concentrations vary between 4.6 and 147.3 Bq/m^3 in different regions with an overall geometric mean of 23.0 Bq/m^3 (GSD 2.61). The geographical distribution pattern shows relatively high inhalation dose rates (>2.0 mSv/y) in the north-eastern part of India. Concentration of uranium and thorium in soil and rocks in northeast areas of India is quiet high. Results of this study showed that Indian dwellings in most areas do not require any action levels with respect to indoor radon due to good natural ventilation in Indian dwellings. Inhalation dose rates on higher side has been observed in the north-eastern part of the country is matter of concern (Pintilie et al. 2018). In another study, Radon (^{222}Rn) in the drinking water of Dehradun City, which is a part of sub Himalayan ranges of India, from the tube wells and hand pumps was measured. Results showed that the values were observed to be more than the average of the recommendations. However, these values were below the highest recommended value of 400 Bq l^{-1}. As study concluded that the water used for drinking in this region is contaminated by radon but still there is lack of big data and more studies are required to explore the study area (Srinivasa et al. 2005).

7.2 Genotoxic Profiling of Radon Gas

Many previous studies are evident (Fig. 8) that radon are capable to inducing genotoxicity in human and other lab animals. Chromosomal aberrations, micronuclei, gene mutations (HPRT), sister chromatid exchanges and DNA damages like markers were noticed in the human peripheral and whole blood lymphocytes, Rat (tracheal epithelial cells), Rabbit (somatic cells), Mouse and rat bone marrow cells (Bilban and Jakopin 2005; Shanahan et al. 1996; Tuschl et al. 1980; Abo-Elmagd et al. 2008; Hornung and Meinhardt 1987).

7.3 Prevention from Radon Exposure (CGR 2012)

1. Testing radon at home is only way to find if people living in these houses are at risk or not. Environmental Protection Agency (EPA) and the surgeon general recommend fixing homes that have levels at or above 4 pCi/L (picocurie per liter). United States have radon programs and provide free radon test kits, which is not very expensive methods to fix and prevent high radon exposure in homes.
2. Sealing visible cracks is a basic part of most radon mitigation approached, but sealing alone is not enough.
3. For dilution of radon, opening of doors and windows may sometimes be effective, but it is not a practical long-term solution.

8 Ionizing Radiation and Cancer

Populations are exposed to radiation from environmental sources, such as nuclear reactor accidents and fallout from weapons testing. In addition, cancer risk occurs from background radiation and ultraviolet radiation. Ionizing radiations induce gene mutations and chromosome aberrations, which are known to be involved in the process of carcinogenesis (Thomas and Symonds 2016). Figure 7 shows different sources of radiation exposures.

Survivors of atomic bomb tragedy in Hiroshima and Nagasaki, who are followed for more than 50 years, provide the evidence based data on the carcinogenic effects of radiation in humans. Epidemiological studies showed that exposure to moderate to high doses of radiation increases the risk of cancer in most organs (Jordan 2016) (Fig. 8).

The Life Span Study (LSS) cohort consists of about 120,000 survivors of the atomic bombings in Hiroshima and Nagasaki, Japan, in 1945. The Radiation Effects Research Foundation (RERF) and its predecessor, the Atomic Bomb Casualty Commission, have studied them. This is one of the longest study with large number of samples and follow up period 1950–2000. Leukaemia was the first cancer linked with

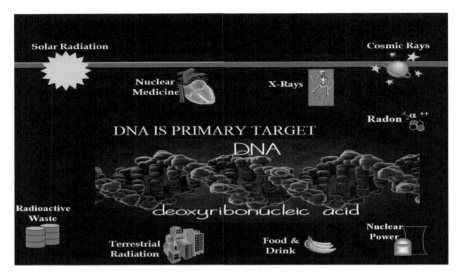

Fig. 7 Various source of radiation increases mutational and cancer risks

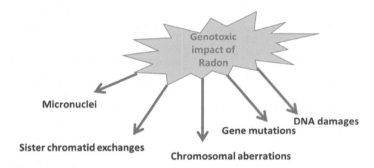

Fig. 8 Genotoxic impact of radon

radiation exposure in atomic bomb survivors (Folley et al. 1952) and has the highest relative risk of any cancer. Results provided by Pierce and colleagues showed that 78 of 176 (44%) leukaemia deaths among survivors with doses exceeding 0.005 Sv were due to radiation exposure (Little 2009).

Cancers of the breast, thyroid and lung risk estimates are fairly precise in addition to leukaemia's, and associations have been found at relatively low doses (<0.2 Gy). Associations between radiation and cancers of the, colon, salivary glands, urinary bladder, ovary, central nervous system and skin have also been reported, but the relationships are not as well quantified. Persons exposed to irradiation for medical reasons have been studied. Medical radiation exposure results in non-uniform doses to the various organs of the body. Therapeutic procedures results in organ specific doses often as high as 40 or more Grey. Association of Leukaemia with medical radiation exposure is documented in many studies. Other medical radiation studies demonstrated a dose-response relationship for breast cancer (Shah et al. 2012). The latency period for radiation exposure cancers like induction of leukaemia is 5–7 years, and for solid tumours is at least 10 years. Radiation exposures by Ultraviolet rays is associated with 65% of melanoma cases and 90% of non-melanoma skin cancers including basal cell carcinoma and squamous cell carcinoma (Testa et al. 2017).

Epidemiological studies provided the "Dose-Response Relationships" for cancer induction following exposure to moderate to high doses of low LET radiation. Although some studies showed radiation effects below 100 mSv but still larger data and more studies are required to yield results which will be statistically significant (Suzuki and Yamashita 2012).

Preventions

(i) Unnecessary exposures as well as over-exposures can be prevented by principle of ALARA "As low as reasonably achievable". Time, Distance, and Shielding are three primary means to keep body safe from radiation exposure (Edison et al. 2017).
Time: Minimize the time of exposure to radiation will decrease the dose.
Distance: The greater the distance between source of radiation and individual, the exposure will be decreased.
Shielding: Shielding is useful for absorbing radiation energy and less of radiation dose is absorbed in the body's tissues. Lead or Lead Equivalent Shielding for X-rays and gamma rays is an effective way to minimize the radiation exposure. When working in radiation areas lead glasses, lead aprons, mobile lead shields, and lead barriers can reduce the exposure to radiations.

(ii) Labeling radioactive and potentially radioactive areas and items will help prevent the spread of contamination.

(iii) Use of personal protective equipment (PPE) such as safety glasses, radiation protection gloves, laboratory coats, thyroid shields are good for prevention. The minimum protective lead equivalents in hand gloves and thyroid shields should be 0.5 mm (Hyun et al. 2016).

(iv) Sign-ages in radiation areas like "no unauthorized entry" or "CONTROLLED RADIATION AREA" with colour recommended by authorities can make public aware of hazard areas.

(v) Education and knowledge about safe handling of all levels of radiation is important to prevent or minimize possible biological effects (Awosan et al. 2016).

In summary, most of the occupational and environmental cancers are preventable. Strong policies should be originated from labour, environment, medical sector, health and other ministries involved in preventing occupational and environmental cancers. Moreover, research in the field of environmental carcinogens can be explored for more solutions.

9 Summary

1. Elimination of risk factors by primary prevention of cancer through mitigation of environmental and occupational determinants.
2. Policies and regulations for phasing out of replaceable processes or substitution of chemicals.
3. Promote safe measures for storage, disposal or recycling of chemicals.
4. Trade and transport of hazardous substances needs strong legislation (e.g. increasing taxes).
5. Public and alternative transportation systems to be utilized like promotion of pedestrian-oriented streets.
6. Standards for Radiation Protection like IAEA/BARC.
7. Education of practitioners to promote the use of referral guidelines.
8. Justification of radiological medical procedures.

Control measures in the working environment

1. Identification and surveillance of exposure
2. Regulating the use of protective equipment for workers
3. Inclusion of occupational cancer in national lists of occupational diseases
4. Notification
5. Establishing national dose registries
6. Providing workers access to information, thus Empowering them

Research in this area

1. Dose response for different types of cancers
2. Low-level exposures to carcinogenic pollutants
3. Environmental risks and genetic susceptibility.

Acknowledgements Authors are grateful to Prof. Sunil Saini, Director Cancer Research Institute and Prof. Mushtaq Ahmad, Dean Himalayan Institute of Medical Sciences, Swami Rama Himalayan University for providing motivation and necessary facilities for writing this book chapter.

References

Abo-Elmagd M, Daif MM, Eissa HM (2008) Cytogenetic effects of radon inhalation. Radiat Meas 43:1265–1269
Andrew TF, Lizbeth LC, Brenda GL, Mariano EC (2017) Standards for arsenic in drinking water: implications for policy in Mexico. J Public Health Policy 38(4):395–406
Awosan KJ, Ibrahim MTO, Saidu SA, Ma'aji SM, Danfulani M, Yunusa EU et al (2016) Knowledge of radiation hazards, radiation protection practices and clinical profile of health workers in a teaching hospital in Northern Nigeria. J Clin Diagn Res 10:LC07–LC12
Bilban M, Jakopin CB (2005) Incidence of cytogenetic damage in lead-zinc mine workers exposed to radon. Mutagenesis 20(3):187–191

Bruce N, Dherani M, Liu R, Hosgood HD, Sapkota Smith KR et al (2010) Does household use of biomass fuel cause lung cancer? A systematic review and evaluation of the evidence for the GBD. Thorax 70:433–441

Burton A (2011) Does the smoke ever really clear? Thirdhand smoke exposure raises new concerns. Environ Health Perspect 119(2):A70–A74

Cao X, MacNaughton P, Laurent JC, Allen JG (2017) Radon-induced lung cancer deaths may be overestimated due to failure to account for confounding by exposure to diesel engine exhaust in BEIR VI miner studies. PLoS ONE 12:e0184298

Cogliano VJ, Baan R, Straif K, Grosse Y, Lauby-Secretan B, Ghissassi FE, Bouvard V et al (2011) Preventable exposures associated with human cancers. J Natl Cancer Inst 103:1827–1839

Diaz OP, Arcos R (2015) Estimation of arsenic intake from drinking water and food (raw and cooked) in a rural village of Northern Chile. Urine as a biomarker of recent exposure. Int J Environ Res Public Health 12:5614–5633

Edison P, Chang PS, Toh GH, Lee LN, Sanamandra SK, Shah VK (2017) Reducing radiation hazard opportunities in neonatal unit: quality improvement in radiation safety practices. BMJ Open Qual 6:e000128

Flanagan SV, Johnston RB, Zheng Y (2012) Arsenic in tube well water in Bangladesh: health and economic impacts and implications for arsenic mitigation. Bull World Health Organ 90:839–846

Folley JH, Borges W, Yamasaki T (1952) Incidence of leukemia in survivors of the atom bomb in Hiroshima and Nagasaki, Japan. Am J Med 13(3):311–321

Hashim D, Boffetta P (2014) Occupational and environmental exposures and cancers in developing countries. Ann Glob Health 80:5393–5411

Hornung R, Meinhardt T (1987) Quantitative risk assessment of lung cancer in U. S. uranium miners. Health Phys 52:417–430

Hyun S, Kim K, Jahng T, Kim H (2016) Efficiency of lead aprons in blocking radiation—how protective are they? Heliyon 2:e00117

IARC: Bray F, Ferlay J, Soerjomataram I, Siegel RL, Torre LA, Jemal A (2018) Department for Management of Noncommunicable Diseases, WHO. Global cancer statistics 2018: GLOBOCAN estimates of incidence and mortality worldwide for 36 cancers in 185, vol 68, pp 394–424

IEPH (International Encyclopedia of Public Health) (2016) Volume 1: 2nd edn

Issaragrisil S, Kaufman DW, Anderson T, Chansung K, Leaverton PE, Shapiro S (2006) The epidemiology of aplastic anemia in Thailand. Blood 107:1299–1307

Jeuland MA, Pattanayak SK (2012) Benefits and costs of improved cookstoves: assessing the implications of variability in health, forest and climate impacts. PLoS One 7:e30338

Jordan BR (2016) The Hiroshima/Nagasaki survivor studies: discrepancies between results and general perception. Genetics 203:1505–1512

Kauts ML, Vink CS, Dzierzak E (2016) Hematopoietic (stem) cell development: how divergent are the roads taken. FEBS Lett 590:3975–3986

Keith S, Doyle JR, Harper C et al (2012) Toxicological profile for radon. Agency for Toxic Substances and Disease Registry (US)

Kim SJ, Williams D, Cheresh P, Kamp DW (2013) Asbestos-induced gastrointestinal cancer: an update. J Gastrointest Dig Syst 3:135

Korte JE, Hertz-Picciotto I, Schulz MR, Ball LM, Duell EJ (2000) The contribution of benzene to smoking-induced leukemia. Environ Health Perspect 108:333–339

Levy BT, Cynthia KW, Niles P, Morehead H, Xu Y, Jeanette MD (2015a) Radon testing: community engagement by a rural family medicine office. J Am Board Fam Med 28:5–10

Levy BT, Wolff CK, Niles P, Morehead H, Xu Y, Daly JM (2015b) Radon testing: community engagement by a rural family medicine. J Am Board Fam Med 28:617–623

Little MP (2009) Cancer and non-cancer effects in Japanese atomic bomb survivors. J Radiol Prot 29:A43–A59

Lodish H, Berk A, Zipursky SL, Matsudaira P, Baltimore D and Darnell J (2000) Proto-oncogenes and tumor-suppressor genes. In: Molecular cell biology, 4th edn. W. H. Freeman, New York

Lugg A, Probert D (1997) Indoor radon gas: a potential health hazard resulting from implementing energy-efficiency measures. Appl Energy 56:93–196

Madaniyazi L, Nagashima T, Guo Y, Yu W, Tong S (2015) Projecting fine particulate matter-related mortality in East China. Environ Sci Technol 49:11141–11150

Markowitz SB, Levin SM, Miller A (2013) Asbestos, asbestosis, smoking, and lung cancer. New findings from the North American insulator cohort. Am J Respir Crit Care Med 188:1–10

Marsili D, Angelini A, Bruno C, Corfiati M, Marinaccio A, Silvestri S et al (2017) Asbestos ban in Italy: a major milestone, not the final cut. Int J Environ Res Public Health 14:1379

McMichael AJ (1988) Carcinogenicity of benzene, toluene and xylene: epidemiological and experimental evidence. IARC Sci Publ 85:3–18

Meharg AA, Rahman MM (2003) Arsenic contamination of Bangladesh paddy field soils: implications for rice contribution to arsenic consumption. Environ Sci Technol 37:229–234

Mukherjee A, Sengupta MK, Hossain MA, Ahamed S, Das B, Nayak B et al (2006) Arsenic contamination in groundwater: a global perspective with emphasis on the Asian scenario. J Health Popul Nutr 24:142–146

Nazaroff WW (2013) Exploring the consequences of climate change for indoor air quality. Environ Res Lett 8:015022

Nielson KK, Rogers VC, Holt RB, Pugh TD, Grondzik WA, de Meijer RJ (1997) Radon penetration of concrete slab cracks, joints, pipe penetrations, and sealants. Health Phys 73:668–678

Oberoi S, Barchowsky A, Wu F (2014) The global burden of disease for skin, lung and bladder cancer caused by arsenic in food. Cancer Epidemiol Biomarkers Prev 23:1187–1194

Ou JY, Fowler B, Ding Q, Kirchhoff AC, Pappas L, Boucher KA et al (2018) Statewide investigation of geographic lung cancer incidence patterns and radon exposure in a low-smoking population. BMC Cancer 18:115

Pintilie V, Ene A, Georgescu L, Moraru DL (2018) Determination of gross alpha, gross beta, and natural radionuclides. Rom J Phys 63:801

Ratnaike RN (2003) Acute and chronic arsenic toxicity. Postgrad Med J 79:391–396

Shah DJ, Sachs RK, Wilson DJ (2012) Radiation-induced cancer: a modern view. Br J Radiol 85:e1166–e1173

Shanahan EM, Peterson D, Roxby D (1996) Mutation rates at the glycophorin A and HPRT loci in uranium miners exposed to radon progeny. Occup Environ Med 53(7):439–444

Sinhaet S, Paul R, Khan I et al (2016) Successful treatment of arsenic-induced lung malignancy: a case report. Int J Med Sci Pub Health 509:1976–1978

Smith AH, Steinmaus CM (2009) Health effects of arsenic and chromium in drinking water: recent human findings. Annu Rev Public Health 30:107–122

Srinivasa E, Rangaswamy DR, Suresh S, Reddy KU, Sannappa J (2005) Measurement of ambient gamma radiation levels and radon concentration in drinking water of Koppa and Narasimharajapura taluks of Chikmagalur district, Karnataka, India. J Radiol Prot 25:475–492

Srogi K (2007) Monitoring of environmental exposure to polycyclic aromatic hydrocarbons: a review. Environ Chem Lett 5:169–195

Suzuki K, Yamashita S (2012) Low-dose radiation exposure and carcinogenesis. Jpn J Clin Oncol 42:563–568

Szeinu J, Beckett WS, Clark N (2000) Guidelines on medical surveillance. Under the occupational safety and health (use and standard of exposure of chemicals hazardous to health) regulations, 2000 P.U.(A)131

Testa U, Castelli G, Pelosi E (2017) Melanoma: genetic abnormalities, tumor progression, clonal evolution and tumor initiating cell. Med Sci (Basel) 5:28

Thomas GA, Symonds P (2016) Radiation exposure and health effects—is it time to reassess the real consequences? R Coll Radiol 28:231–236

Tuschl H, Altmann H, Kovac R (1980) Effects of low-dose radiation on repair processes in human lymphocytes. Radiat Res 81:1–9

Environmental Toxicants and Male Reproductive Toxicity: Oxidation-Reduction Potential as a New Marker of Oxidative Stress in Infertile Men

Shubhadeep Roychoudhury, Manas Ranjan Saha and Mriganka Mouli Saha

Abstract Exposure to various environmental and lifestyle-dependent factors such as heavy and trace metals, hydrocarbons, ethylene glycol ethers, obesity, tobacco, alcohol and recreational drugs etc. have been identified to cause reproductive toxicity in men. A number of toxicants affect spermatogenesis leading to poor semen quality affecting fertility in such men, primarily through the mechanism of oxidative stress. In the male reproductive system, oxidative stress is brought about either by excessive production of extrinsic free radicals or by reduced activity of intrinsic antioxidants thereby disrupting the redox balance. Discrete measures of reactive oxygen species, total antioxidant capacity, and post hoc damage suggest an ambiguous relationship between the redox system and male fertility. Antioxidants work by donating electrons to the oxidants, thereby reducing the chances of oxidants to acquire electrons from other nearby structures and cause oxidative damage. Oxidation-reduction potential (ORP) measures this relationship between oxidants and antioxidants in semen. The MiOXSYS system used to measure ORP requires a small volume (~30 µl) of liquefied semen and the measurement is completed in less than 5 min. The galvanostat-based analyzer uses electrochemical technology to measure the ORP in millivolts (mV) which is then normalized to express as $mV/10^6$ sperm/mL. The role of ORP as a surrogate marker to conventional semen quality parameters is a current topic of investigation by a number of researchers and clinicians. It can be measured in semen and seminal plasma up to 2 h of liquefaction. ORP correlates negatively with conventional as well as advanced semen quality parameters, including sperm concentration, total sperm count, total motile sperm count, motility, morphology, and DNA fragmentation thus confirming the association of oxidative stress with male factor infertility. ORP values can differentiate the degree of oxidative stress-induced male

S. Roychoudhury (✉)
Department of Life Science and Bioinformatics, Assam University, Silchar 788011, India
e-mail: shubhadeep1@gmail.com

M. R. Saha
University of North Bengal, Raja Rammohunpur 734 013, West Bengal, India

M. M. Saha
Department of Obstetrics and Gynaecology, College of Medicine and J. N. M. Hospital, Kalyani 741 235, West Bengal, India

© Springer Nature Switzerland AG 2019
K. K. Kesari (ed.), *Networking of Mutagens in Environmental Toxicology*, Environmental Science,
https://doi.org/10.1007/978-3-319-96511-6_5

infertility. A number of clinical studies involving cohorts of men from USA, Qatar and India have established seminal ORP cut-off values to distinguish fertile men from infertile patients. Monitoring ORP levels may help predict treatment efficacy in patients as higher ORP values are indicative of the progression of infertility. It can also be measured in cryopreserved semen samples, which is important in predicting the success of assisted reproductive techniques (ART). A recent ART study reported higher clinical pregnancy rate in infertile men with low seminal ORP in comparison to patients with high ORP. Findings of recent clinical investigations indicate ORP as a novel, independent and robust diagnostic marker of seminal oxidative stress that should find its place in the male infertility workup algorithm.

Keywords Environment · Lifestyle · Toxicity · Semen · Male reproduction · Infertility · Oxidative stress · Oxidation-reduction potential

1 Introduction

Reproduction is a natural process for most of the couples involving neither special planning nor intervention. However, 15% of couples struggle to conceive after one year of regular, unprotected intercourse and, consequently, seek medical advice on how to improve their chances of fertilization and successful pregnancy (Trussell 2013). Hence, infertility has become the most important public health concern affecting 48.50 million couples globally (Agarwal et al. 2015a, b) wherein only the male factor accounts for 40–50% of infertility cases (Kumar and Singh 2015). Several toxicants of chemical and physical origin generated by industrial and agricultural activities are released into the environment constitute a putative hazard to the fertility of men (Spira and Multigner 1998). Exposure to various environmental and lifestyle-dependent factors have been associated with excessive production of extrinsic free radicals or reduced activity of intrinsic antioxidants thereby causing oxidative stress in the male reproductive system that may gradually manifest into reproductive toxicity affecting fertility in such men (Jana and Sen 2012; Pizent et al. 2012; Aitken et al. 2014; Gabrielsen and Tanrikut 2016). Clinicians largely rely on routine semen analysis for the diagnosis of male infertility in spite of poor association of conventional semen parameters with male fertility potential (Björndahl et al. 2015; Agarwal et al. 2017a). However, it is felt by many clinicians as well as researchers that male fertility evaluation should not be based on conventional semen analysis alone and more reliable, quantifiable, unbiased and universal functional measures of semen quality must be incorporated in male infertility evaluation (Esteves 2014; Agarwal et al. 2017a).

2 Environmental Toxicants and Male Reproductive Functions

A number of environmental toxicants have been identified to affect spermatogenesis leading to low sperm count, abnormal sperm morphology and eventually poor semen quality. Various classes of compounds such as heavy metals, organic polychlorinated dibenzodioxins, dicarboximide fungicides, environmental phenols and several other different classes of pollutants and chemicals are often released into the environment during industrial processes which are gradually taken up by humans through the ingestion of contaminated food and water, usage of consumer products (e.g. plasticware and cosmetics etc.), inhalation of polluted air and so on (Spira and Multigner 1998; Aitken et al. 2004; Sharpe 2010).

2.1 Heavy and Trace Metals

Plethora of evidences revealed negative impact of heavy metals including cadmium, lead, manganese, chromium, copper, mercury, nickel and silver on male infertility. Significantly higher levels of cadmium was reported in blood and seminal plasma of infertile patients in comparison to fertile men or men from the normal population upon environmental exposure which was further validated in animal model and reduced sperm concentration and motility was noted (Benoff et al. 2009). Seminal plasma cadmium level was found to be significantly higher than the serum cadmium level when it was tested among 60 infertile males and 40 normozoospermic subjects (Akinloye et al. 2006). Application of low dose of cadmium was able to disrupt inter-Sertoli cell tight junctions in rats leading to disruption of spermatogenesis (Siu et al. 2009). Cadmium also contributes to infertility in males with varicoceles as the percentage of apoptotic nuclei and testicular cadmium levels were found to be high in such men (Benoff et al. 2004). Lead is another heavy metal that passes to the bloodstream and is incorporated into the tissues, including hypothalamus, hypophysis, and testes due to ingestion or inhalation, affecting neuroendocrine system (Lamb and Bennett 1994). Reduced sperm cell formation was observed in 150 workers exposed to lead in their workplaces (Lancranjan et al. 1975). High lead concentration in blood was also noted among people working in batteries and paint factories with decreased sperm velocity, reduced sperm motility suggesting retarded sperm activity (Lamb and Bennett 1994; Naha and Chowdhury 2006). Literatures revealed that manganese could alter reproductive functions, elevated levels exhibiting harmful effects on sperm morphology and motility (Li et al. 2012). Copper, an essential trace metal, can decrease sperm function including concentration, viability and motility in mammalian model in its ionic and non-ionic forms in a dose-dependent and time-dependent manner indicating the possibility of its adverse affect on male fertility (Holland and White 1988; Roychoudhury and Massanyi 2008; Roychoudhury et al. 2010, 2016a, b). Molybdenum concentration was found to be the most consistent in blood with reduced sperm concentration and morphology when a bunch of essential and nonessential metals (arsenic, cadmium, chromium, copper, lead, manganese, mercury, molybdenum, selenium and zinc) were subjected to assess their effects on semen quality (Meeker et al. 2008).

2.2 Pollutants

Several pollutants are associated with deteriorating seminal quality depending upon the dose and time of exposure. Workers in tollgates were found to have lower sperm motility including lower progressive motility and sperm kinetics than the control males (Rosa et al. 2003). Hydrocarbons such as toluene, benzene and xylene were reported in the blood and semen of some workers at workplaces where their air concentration exceeded the maximum permissible levels resulting in decreased sperm vitality and motility in the occupationally exposed men (Xiao et al. 2001). Adverse effect of dioxin exposure during infancy or at puberty was associated with reduction in sperm concentration, progressive motility, total motile sperm count, estradiol and an increase in follicle stimulating hormone (Mocarelli et al. 2008). It is also believed that environmental pesticide exposures can adversely affect spermatogenesis in men at large. Environmental exposures to polychlorinated biphenyls (PCB) and dichlorodiphenyldichloroethylene (DDE) were found to be associated with altered semen quality parameters when a cross-sectional study was conducted on 212 male partners of subfertile couples (Hauser et al. 2003).

2.3 Chemicals

Plethora of evidence reflects the pronounced adverse effect of chemical exposures during adulthood affecting testicular and post-testicular functions and male fertility. For instance, ethylene glycol ethers produced by chemical industry (especially, ethylene glycol methyl ether and ethylene glycol ethyl ether) exerts deleterious effects on reproduction and fertility in mammalian models (Boatman 2001; Multigner et al. 2005). In a human study, exposure to glycol ether was linked to a secular decrease in semen quality (Multigner et al. 2007). Another common industrial chemical exhaust, bisphenol A (BPA) has been associated with lowered sperm count and motility in men who worked in the BPA-based factories (Rahman et al. 2015; Manguez-Alarcon et al. 2016).

3 Lifestyle and Male Reproductive Functions

There are some lifestyle-dependent factors including obesity, smoking, alcohol and drugs that play a vital role in male reproductive health. However, these factors reflect less conclusive evidences affecting semen quality and male fertility.

3.1 Obesity

Obesity is an important lifestyle-dependant factor that has negative impact on spermatogenesis and/or male fertility. Evidences suggest that men with poor semen quality are three times more likely to be obese than men with normal semen quality. Male infertility was found to be associated with a higher incidence of obesity exhibiting reduced androgen levels and sex hormone-binding globulin levels accompanied by elevated estrogen levels, thereby indicating endocrine dysregulation in obese men and increased risk of altered semen parameters and infertility (Magnusdottir et al. 2005; Hammoud et al. 2008). In overweight and obese men (BMI $\geq$ 25 kg/m^2) mean sperm concentration was found to be lower than those of normal-weight men (BMI 20–25 kg/m^2) (Jensen et al. 2004). Obesity was also associated with a 1.3-fold relative risk for erectile dysfunction which can be explained by the decreased T levels and elevated levels of several pro-inflammatory cytokines in such individuals (Bacon et al. 2003; Seftel 2006).

3.2 Smoking and Alcohol

Cigarette smoking and alcohol consumption are two major recreational factors that usually come top of the list of affecting male reproductive health. Sperm density was reported to be much lower among smokers in comparison to non-smokers (Vine et al. 1996). Among smokers, reduced sperm density, decreased total sperm count and lower total number of motile sperm was observed, too (Kunzle et al. 2003). Another study strongly associated tobacco chewing males in India with a decrease in their semen quality and the extent of oligoasthenozoospermia or azoospermia (Said et al. 2005). Smoking was believed to be the cause of seminal oxidative stress, lack of sperm plasma membrane integrity and DNA fragmentation that directly corroborate with male infertility (Saleh et al. 2002; Belcheva et al. 2004; Sepaniak et al. 2006). Alcohol consumption has been associated with impairment of spermatogenesis and reduction in sperm counts and testosterone levels (Muthusami and Chinnaswamy 2005). A significant decrease in the number of sperm cells was noted due to severe alcohol intake (Donnelly et al. 1999; La Vignera et al. 2013). A synergistic effect of cigarette smoking and alcohol consumption results in significant reduction of seminal volume, sperm concentration, percentage of motile spermatozoa, and increased number of non-motile viable gametes (Martini et al. 2004; Guthauser et al. 2013).

3.3 Recreational Drugs

A few studies demonstrated the adverse impact of recreational drugs such as cocaine, cannabis and marijuana on male infertility although the actual mechanism is not very clear. Use of cocaine for five or more years was found to be associated with lower sperm motility, concentration and abnormal morphology (Bracken et al. 1990). Similar results were also observed in chronic use of marijuana (Harclerode 1984). Deleterious effects of delta-9-tetrahydrocannabinol, the active compound of marijuana on sperm function was evident from the reduction in sperm progressive motility and decreased acrosome reaction (Whan et al. 2006). In addition, methadone and heroin were found to cause lower serum testosterone concentrations, reduced sperm motility, lower ejaculate volumes and abnormal sexual dysfunction (Fronczak et al. 2012). Use of anabolic androgen steroids by athletes, weightlifters and bodybuilders induces a state of hypogonadotrophic hypogonadism by means of decreasing testosterone concentration thereby reducing spermatogenesis (Knuth et al. 1989; Karila et al. 2004). Interestingly, a few prescribed drugs of different types may also exert adverse impact on spermatogenesis. For instance, sulfasalazine, used for the treatment of irritable bowel disorders and some other chemotherapeutic agents (e.g. cyclophosphamide) used for treatment of cancers or kidney diseases may induce infertility in men (Nudell et al. 2002; Feagins and Kane 2009; Semet et al. 2017).

4 Oxidative Stress, Male Reproductive Toxicity and Infertility

Reactive oxygen species (ROS) are molecules that contain an oxygen atom with an unpaired electron in outer shell. Due to the unpaired electron in the outermost shell they become very unstable and these unstable forms of oxygen are called free radicals. ROS are produced in living cells either from intrinsic or extrinsic sources as byproducts of cellular metabolism resulting from mitochondrial respiration through oxidative phosphorylation (Sharma and Agarwal 1996; Dickinson and Chang 2011; Chen et al. 2003; Lavranos et al. 2012; Roychoudhury et al. 2017a, b). ROS hinder the cells' own natural antioxidant defense system. Endogenous antioxidants or those acquired from diet limit the damage to cells by detoxifying these reactive intermediates by donating an electron to ROS to stabilize them (Finkel 2011). Under normal physiological circumstances ROS are present in genital tract in low concentration (Guerin et al. 2001; Roychoudhury et al. 2017a, b) which is necessary for functions of the male gamete including capacitation, hyperactivation and acrosome reaction

(Aitken and Fisher 1994; Sharma and Agarwal 1996). Oxidative stress occurs when the delicate balance in the redox system that maintains equilibrium between the ROS and the antioxidants, is disturbed (Agarwal et al. 2017a; Roychoudhury et al. 2017a, b). Studies indicated that infertile men are more likely to have higher concentrations of ROS and lower level of antioxidants in their seminal plasma (de Lamirande and Gagnon 1995; Roychoudhury et al. 2016a, b). ROS causes reproductive toxicity and impairs fertility by means of two principal mechanisms. First, being rich in polyun-saturated fatty acids, sperm cell membranes are highly vulnerable to ROS; thereby lipid peroxidation of the plasma membrane takes place which in turn reduces sperm motility, quality and fertility (Henkel 2011; Aitken et al. 2014, 2016). Secondly, ROS directly damage sperm DNA affecting the purine and pyrimidine bases and the deoxyribose backbone (Agarwal et al. 2003; Oliva 2006). Additionally, ROS may initiate apoptosis within the sperm, leading to caspase-mediated enzymatic degra-dation of the sperm DNA (Moustafa et al. 2004; Villegas et al. 2005). Leukocytes especially neutrophils and macrophages, as well as immature spermatozoa include the major endogenous sources of ROS. Environmental and lifestyle factors discussed earlier in this chapter play an important role in production of exogenous ROS leading to male reproductive toxicity and infertility which have been confirmed by studies in human subjects as well as mammalian models (Saleh et al. 2003; Gharagozloo and Aitken 2011).

Amongst several environmental factors mentioned above, cadmium directly induces oxidative stress in testes affecting sperm production, testosterone activity, secretion, spermatogenesis impairment as well as it decreases antioxidant defense system by means of reducing superoxide oxidase, glutathione peroxidase and cata-lase enzyme activity of seminal plasma membrane (Turner and Lysiak 2008; Sarkar et al. 2013). Testicular cadmium level has been associated with varicocele which is the most common correctable cause of male infertility and oxidative stress appears to be the primary mechanism of varicocele-induced injury (Benoff et al. 2004; Jensen et al. 2017). Lead exposure associated with oxidative stress contributes to an imbal-ance in the reproductive system disrupting testicular steroidogenesis by inhibiting the activities of testicular steroidogenic enzymes (Liu et al. 2008). Lead toxicity is also manifested by its deposition in testes, epididymis, vas deferens, seminal vesicle and seminal ejaculate which in turn reduce the sperm count, motility and germ cell pop-ulation (Adhikari et al. 2001; Chowdhury 2009). Copper facilitates the production of superoxide radicals, hydroxyl radicals and hydrogen peroxide via the Haber-Weiss reaction causing oxidative damage and initiate adverse effect on spermatozoa con-centration, viability and motility in animal models including reproductive toxicity of copper oxide nanoparticles (Roychoudhury and Massanyi 2008; Roychoudhury et al. 2010, 2016a, b). Manganese, required as a cofactor in many cellular enzymes includ-ing arginase, superoxide dismutase, alkaline phosphatase etc. has become a global concern due to its increased release into the environment which in turn accumulates in mitochondria, disrupting oxidative phosphorylation and increases the generation of ROS (Gunter et al. 2006). Manganese intoxication has been reported to lower synthesis and secretion of testosterone by acting directly on the Leydig cells or indi-rectly by acting on the anterior pituitary gland inhibiting the secretion of luteinizing hormone which in turn inhibits androgen biosynthesis in Leydig cells (Chandel and Jain 2017).

Formaldehyde, a ubiquitous environmental pollutant, exerts detrimental effects on the reproduction, respiratory and haematological systems by means of production of excessive ROS (Zhou et al. 2006). Studies revealed that use of formaldehyde can lead to testicular atrophy and decreased testes weight, diameter of seminiferous tubules, seminiferous epithelial height and decreased number of spermatozoids (Golalipour et al. 2007; Gules and Eren 2010). Inhalation of toluene alters the hormonal status of the anterior pituitary gland in rodents as well as it facilitates oxidative damage to DNA and reproductive toxicity by means of decreased sperm number and increased 8-oxo-2′-deoxyguanosine formation in sperm cells of the testis (Nakai et al. 2003). Similarly, toxicity of benzene may result from oxidative metabolism of benzene to reactive products which ultimately cause DNA damage and this could be the possible mechanism by which benzene acts as a toxicant for spermatogenesis (Song et al. 2005). Long term exposure to xylene leads to reproductive toxicity through decreased spermatozoa viability, decreased motility with lower acrosin action from spermatozoa (Xiao et al. 2001). Polychlorinated biphenyls are the most environmentally persistent pollutants that disrupt the endocrine system (Apostoli et al. 2003) by reducing testosterone synthesis and steroidogenic enzyme activity in Leydig as well as Sertoli cells (Fiandanese et al. 2016), while ROS-induced BPA concentration is negatively associated with sperm concentration, normal morphology, and sperm DNA damage (Meeker et al. 2011). Administration of ethylene glycol monoethyl ether was found to be adversely affect steroidogenesis in rodents by decreasing the expression of steroid acute regulatory protein and androgen-binding protein (Adedara and Farombi 2013).

Furthermore, lifestyle-depended factors, such as obesity has been shown to be associated with the production of ROS which results in decreased sperm density and total count and significant negative correlation to increasing body mass index dysregulating the action of hypothalamic-pituitary-gonadal axis (Furukawa et al. 2004). Smoking and marijuana inhalation have been the key factors in the production of excessive ROS, which help in pathogenesis of several diseases including reproductive toxicity thereby inhibiting sperm motility, viability and thus fertility of the male (Close et al. 1990; Whan et al. 2006; La Maestra et al. 2015). On the other hand, although tobacco chewing is comparatively less harmful than smoking, it is not harmless altogether. It increases the risk of multiple oral premalignant lesions and affects semen parameters in a dose-dependent manner including reduced sperm concentration, motility, morphology, and viability (Said et al. 2005; Sunanda et al. 2014). However, the actual mechanism is still not known.

From the above discussion, it is clear that several environmental as well as lifestyle-dependent factors are involved in the pathophysiology of male infertility by means of mechanisms including oxidative stress-induced reproductive toxicity. Excessive increase in the generation of ROS and/or excessive decrease in the level of antioxidants may disrupt the seminal redox balance and trigger molecular changes that induce deterioration of semen quality and associated male fertility parameters including oxidative damage to sperm DNA, lipids and proteins. Therefore, alleviation of oxidative stress constitutes a potential treatment strategy for male infertility (Agarwal et al. 2017a). Various direct and indirect tests used by different laboratories for measurement of seminal oxidative stress are discussed below.

5 Measurement of Oxidative Stress in Semen

Seminal oxidative stress is commonly measured by means of quantifying the ROS via chemiluminescence assay or by measuring the total antioxidant capacity (TAC assay) or post hoc damage by malondialdehyde (MDA) assay. The chemiluminescent ROS assay is based on the reaction between luminol and oxidizing compounds distinguishing poor quality of semen samples from good quality ones (Agarwal et al. 2014a, 2015a, b). The TAC assay is based on the ability of antioxidants present in semen to scavenge the ROS through specific or non-specific mechanisms (Muller et al. 2013). Amongst several TAC assays, Trolox equivalent antioxidant capacity is the most widely accepted test for determining seminal TAC levels between poor and good quality semen samples in order to differentiate infertile men from healthy ones (Roychoudhury et al. 2016a, b). The MDA assay determines the damage done to proximate lipids by free radicals in semen samples (Marnett 1999). However, these conventional approaches are single marker measurements that fail to capture both of the components of oxidative stress i.e. oxidants, and antioxidants/reductants (Agarwal et al. 2015a, b, 2017a). All of these have their own drawbacks as such assays are tedious, time consuming, involve sophisticated instruments and/or require special technical skills and large semen volumes causing difficulty in providing the full picture of the true oxidative state of the sample. Hence, a comprehensive measure of the activity of all known as well as unknown oxidants and antioxidants in a semen sample will better describe the state of the redox system thus facilitating better diagnosis and treatment of male infertility by the clinician.

6 Oxidation-Reduction Potential of Semen: A New Diagnostic Marker in Male Infertility

Antioxidants work by donating electrons to the oxidants, thereby reducing the chances of oxidants to acquire electrons from other nearby structures and cause oxidative damage. Oxidation-reduction potential (ORP) measures this relationship between oxidants and antioxidants in fluids including semen. Validation of a novel ORP diagnostic platform by comparison to mass spectrometry using disposable electrodes has paved the way for a rapid and comprehensive status of redox state in a sample. The difference between oxidants and antioxidants (reductants) is detected as electrical signal produced by oxidation of an electrode under standardized conditions without determining contributions of individual molecules involved (Roychoudhury et al. 2017a, b; Agarwal et al. 2018; Polson et al. 2018). Based on the electrochemical technology, the MiOXSYS system uses a platinum-based electrode sensor with an Ag/AgCl reference cell, and a galvanostat-based analyzer, which completes the circuit. A small volume (~30 µl) of liquefied semen sample is added to the pre-inserted sensor and allowed to flow across the working electrode and to fill the reference cell, thereby completing the electrochemical circuit. Voltage is measured between the reference cell and working electrode every 0.5 s (or 2 Hz), while the counter is set to a voltage sufficient to achieve a 1 nA stabilizing current. The resulting ORP measurement is displayed in millivolts (mV) reflecting a net average of the run (Roychoudhury et al. 2017a, b). The entire process takes less than 5 min. The raw ORP value displayed by the analyzer (mV) is divided by the sperm concentration (sperm count $\times$ 10^6/mL) to obtain the normalized ORP, which is expressed as mV/10^6 sperm/mL (Agarwal et al. 2017a).

In the diagnosis of male infertility, the role of ORP as a surrogate marker to conventional semen quality parameters is a current topic of investigation by a number of researchers and clinicians. It facilitates wider application of oxidative stress measurement in clinical and research settings. ORP can be measured in neat semen and seminal plasma and the measurements are not affected by the age of semen or seminal plasma for up to 2 h of liquefaction (Agarwal et al. 2016b). ORP correlates negatively with conventional as well as advanced semen quality parameters, such as sperm concentration (Agarwal et al. 2016b, 2017a; Agarwal and Wang 2017; Roychoudhury et al. 2017a, b; Toor et al. 2016), total sperm count (Agarwal et al. 2016b, 2017a; Toor et al. 2016), total motile sperm count (Al Said et al. 2017), motility (Agarwal et al. 2017a; Agarwal and Wang 2017; Toor et al. 2016), morphology (Roychoudhury et al. 2017a, b; Majzoub et al. 2017), and DNA fragmentation (Arafa et al. 2017) confirming the association of oxidative stress with poor semen quality.

ORP values can differentiate the degree of oxidative stress-induced male factor infertility. Using a cohort of fertile and infertile men from USA a seminal ORP cutoff value 1.36 mV/10^6 sperm/mL was established for distinguishing fertile men from infertile patients (Agarwal et al. 2017a). A couple of studies conducted collectively and individually between Cleveland Clinic (USA) and Doha (Qatar) recommended similar ORP cutoff values 1.41 and 1.42 mV/10^6 sperm/mL to distinguish fertile from infertile men (Agarwal et al. 2017b; Arafa et al. 2018). Recently, from a cohort of fertile and infertile men in India, Roychoudhury et al. (2017a, b) established a seminal ORP cutoff 1.23 mV/10^6 sperm/mL to distinguish healthy men from infertile patients. Furthermore, ORP is highly predictive of oligozoospermia and asthenozoospermia (Agarwal et al. 2017b), and an ORP cutoff 2.59 mV/10^6 sperm/mL best predicted oligozoospermia using a cohort of men from USA (Agarwal and Wang 2017). Monitoring seminal ORP levels over time may help predict the efficacy of antioxidant therapies and define effective doses and durations of treatment. Monitoring ORP as a marker of oxidative stress has also been proposed in cases of leukocytospermia because ORP values paralleled the levels of biomarkers of active inflammation (Hagan et al. 2015; Sikka et al. 2016).

In the human body, the sperm cells live in an aerobic environment like all other cells, and hence, are exposed to different redox states depending on the prevailing circumstances (Naviaux 2012; Agarwal et al. 2016a, b). ORP can also be measured in cryopreserved semen samples, which is important in predicting the success of assisted reproductive techniques (ART) (Agarwal et al. 2016b). In ART procedures sperm cells are exposed continuously to several culture media and incubating conditions starting from sperm preparation to sperm cryopreservation. An optimum redox potential is required for successful embryogenesis and to avoid teratogenesis (Ufer et al. 2010), and the ORP of the culture medium might be involved in the regulation of the fertilization process (Panner Selvam et al. 2018). It is important to maintain the ORP of culture media on the lower side in order to neutralize and counteract ROS produced especially during the centrifugation process of abnormal semen samples from infertile men (Agarwal et al. 2014b). Determination of ORP values of 10 different culture media commonly used for sperm preparation and ART revealed lower values of the sequential culture medium and one-step culture medium compared to the sperm wash media. SAGE-1-Step medium recorded the lowest ORP value of 208.63 mV (Panner Selvam et al. 2018). In intracytoplasmic sperm injection (ICSI), during sperm processing seminal plasma (that contains protective antioxidants) is removed keeping the sperm cell vulnerable to possible attack by toxic oxygen metabolites generated by immature spermatozoa and leucocytes (Aitken and Baker 1995; Zini et al. 2009). In ICSI procedures, post-washed sperm specimens are loaded into a microdroplet containing a viscous medium of polyvinylpyrrolidone or hyaluronic acid that slowdown the sperm movement thereby facilitating sperm selection, handling and immobilization (Roychoudhury et al. 2018). Comparison of ORP levels of sperm cells exposed to polyvinylpyrrolidone and hyaluronic acid in an experimental ICSI study suggested lower ORP levels in polyvinylpyrrolidone-selected sperm than using hyaluronic acid after 20 min and 1 h of exposure. This indicated that the lower levels of oxidative stress are found in washed sperm cells selected in the polyvinylpyrrolidone-based medium (Roychoudhury et al. 2018). In another clinical study on embryo quality and clinical pregnancy rate, patients with low seminal ORP (<1.36 mV/10^6 sperm/mL) had higher clinical pregnancy rate in comparison to the group with high ORP (Ayaz et al. 2017).

Discrete measures of free radicals, antioxidant activity, and oxidative damage suggest an ambiguous relationship between the redox system and male fertility. Measuring ORP can help rule in male infertility cases associated with oxidative stress that would otherwise go undetected with a conventional semen analysis. Data generated from recent clinical and experimental studies poses ORP as a novel, independent and robust diagnostic marker of male infertility.

References

Adedara IA, Farombi EO (2013) Chemoprotective effects of kolaviron on ethylene glycol monoethyl ether-induced pituitary-thyroid axis toxicity in male rats. Andrologia 45(2):111–119

Adhikari N, Sinha N, Narayan R, Saxena DK (2001) Lead-induced cell death in testes of young rats. J Appl Toxicol 21:275–277

Agarwal A, Wang SM (2017) Clinical relevance of oxidation-reduction potential in the evaluation of male infertility. Urology 104:84–89

Agarwal A, Saleh RA, Bedaiwy MA (2003) Role of reactive oxygen species in the pathophysiology of human reproduction. Fertil Steril 79:829–843

Agarwal A, Tvrda E, Sharma R (2014a) Relationship amongst teratozoospermia, seminal oxidative stress and male infertility. Reprod Biol Endocrinol 12:45

Agarwal A, Durairajanayagam D, Virk G, du Plessis SS (2014b) Sources of ROS in ART. In: Agarwal A, Durairajanayagam D, Virk G, du Plessis SS (eds) Strategies to ameliorate oxidative stress during assisted reproduction. Springer, Cham, Switzerland, pp 3–22

Agarwal A, Mulgund A, Hamada A, Chyatte M (2015a) A unique view on male infertility around the globe. Reprod Biol Endocrinol 13:37

Agarwal A, Ahmad G, Sharma R (2015b) Reference values of reactive oxygen species in seminal ejaculates using chemiluminescence assay. J Assist Reprod Genet 32:1721–1729

Agarwal A, Roychoudhury S, Bjugstad KB, Cho CL (2016a) Oxidation-reduction potential of semen: what is its role in the treatment of male infertility? Ther Adv Urol 8:302–318

Agarwal A, Sharma R, Roychoudhury S, du Plessis S, Sabanegh E (2016b) MiOXSYS: a novel method of measuring oxidation-reduction potential in semen and seminal plasma. Fertil Steril 106:566–573

Agarwal A, Roychoudhury S, Sharma R, Gupta S, Majzoub A, Sabanegh E (2017a) Diagnostic application of oxidation-reduction potential assay for measurement of oxidative stress: clinical utility in male factor infertility. Reprod Biomed Online 34:48–57

Agarwal A, Arafa M, Chandrakumar R, Majzoub A, AlSaid S, Elbardisi H (2017b) A multicenter study to evaluate oxidative stress by oxidation-reduction potential, a reliable and reproducible method. Andrology 5:939–945

Agarwal A, Henkel R, Sharma R, Tadros N, Sabanegh E (2018) Determination of seminal oxidation-reduction potential (ORP) as an easy and cost-effective clinical marker of male infertility. Andrologia 50(3):Epub 2017 Oct 23

Aitken RJ, Baker HW (1995) Seminal leukocytes: passengers, terrorists or good samaritans? Hum Reprod 10:1736–1739

Aitken J, Fisher H (1994) Reactive oxygen species generation and human spermatozoa: the balance of benefit and risk. Bioessays 16(4):259–267

Aitken RJ, Koopman P, Lewis SEM (2004) Seeds of concern. Nature 432:48–52

Aitken RJ, Smith TB, Jobling MS, Baker MA, De Iuliis GN (2014) Oxidative stress and male reproductive health. Asian J Androl 16:31–38

Aitken RJ, Gibb Z, Baker MA, Drever J, Gharagozloo P (2016) Causes and consequences of oxidative stress in spermatozoa. Reprod Fertil Dev 28:1–10

Akinloye O, Arowojolu AO, Shittu OB, Anetor JI (2006) Cadmium toxicity: a possible cause of male infertility in Nigeria. Reprod Biol 6:17–30

Al Said S, Majzoub A, Arafa M, El Bardisi H, Agarwal A, Al Rumaihi K (2017) Oxidation-reduction potential: a valuable tool for male fertility evaluation. In: Poster presented at: 33rd annual european society for health and reproduction conference, July 3–5, Geneva, Switzerland

Apostoli P, Mangili A, Carasi S, Manno M (2003) Relationship between PCBs in blood and D-glucaric acid in urine. Toxicol Lett 144(1):17–26

Arafa M, ElBardisi H, Majzoub A, Al Said S, Al Nawasra H, Khalafalla K et al (2017) Correlation of sperm DNA fragmentation and seminal oxidation-reduction potential in infertile men. In: Poster presented at: 33rd annual European society for health and reproduction conference, July 3–5, Geneva, Switzerland

Arafa M, Agarwal A, Al Said S, Majzoub A, Sharma R, Bjugstad KB, Al Rumaihi K, Elbardisi H (2018) Semen quality and infertility status can be identified through measures of oxidation-reduction potential. Andrologia 50(2):Epub 2017 Aug 3

Ayaz A, Balaban B, Sikka S, Isiklar A, Tasdemir M, Urman B (2017) Effect of seminal ORP value on embryo quality and clinical pregnancy rate. In: Poster presented at: 33rd annual European society for health and reproduction conference, 2017 July 3–5, Geneva, Switzerland

Bacon CG, Mittleman MA, Kawachi I, Giovannucci E, Glasser DB, Rimm EB (2003) Sexual function in men older than 50 years of age: results from the health professionals follow-up study. Ann Intern Med 139:161–168

Belcheva A, Ivanova-Kicheva M, Tzvetkova P, Marinov M (2004) Effects of cigarette smoking on sperm plasma membrane integrity and DNA fragmentation. Int J Androl 27:296–300

Benoff SH, Milan C, Hurley IR, Napolitano B, Marmar JL (2004) Bilateral increased apoptosis and bilateral accumulation of cadmium in infertile men with left varicocele. Hum Reprod 19(3):616–627

Benoff S, Hauser R, Marmar JL, Harley IR, Napolitano B, Centola GM (2009) Cadmium concentrations in blood and seminal plasma: correlations with sperm number and motility in three male populations (infertility patients, artificial insemination donors, and unselected volunteers). Mol Med 15(7–8):248–262

Björndahl L, Barratt CL, Mortimer D, Jouannet P (2015) 'How to count sperm properly': checklist for acceptability of studies based on human semen analysis. Hum Reprod. dev305

Boatman RJ (2001) Glycol ethers: ethers of propylene, butylene glycols, and other glycol derivatives. In: Cohrssen B, Bingham E, Powell CH (eds) Patty's toxicology. Wiley

Bracken MB, Eskenazi B, Sachse K, McSharry JE, Hellenbrand K, Leo-Summers L (1990) Association of cocaine use with sperm concentration, motility, and morphology. Fertil Steril 53:315–322

Chandel M, Jain GC (2017) Effects of manganese exposure on testicular histomorphology, serum hormones level and biochemical marker parameters in Wistar rats. J Biol Sci Med 3(3):1–14

Chen Q, Vazquez EJ, Moghaddas S et al (2003) Production of reactive oxygen species by mitochondria: central role of complex III. J Biol Chem 278(38):36027–36031

Chowdhury AR (2009) Recent advances in heavy metals induced effect on male reproductive function-A retrospective. Al Ameen J Med Sci 2(2):37–42

Close CE, Roberts PL, Berger RE (1990) Cigarettes, alcohol and marijuana are related to pyospermia in infertile men. J Urol 144:900–903

de Lamirande E, Gagnon C (1995) Impact of reactive oxygen species on spermatozoa: a balancing act between beneficial and detrimental effects. Hum Reprod 10(Suppl 1):15–21

Dickinson BC, Chang CJ (2011) Chemistry and biology of reactive oxygen species in signaling or stress responses. Nat Chem Biol 7(8):504–511

Donnelly GP, McClure N, Kennedy MS, Lewis SEM (1999) Direct effect of alcohol on the motility and morphology of human spermatozoa. Andrologia 31:43–47

Esteves S (2014) Clinical relevance of routine semen analysis and controversies surrounding the 2010 World Health Organization criteria for semen examination. Int Braz J Urol 40:443–453

Feagins LA, Kane SV (2009) Sexual and reproductive issues for men with inflammatory bowel disease. Am J Gastroenterol 104:768–773

Fiandanese N, Borromeo V, Berrini A, Fischer B, Schaedlich K, Schmidt JS, Secchi C, Pocar P (2016) Maternal exposure to a mixture of di(2-ethylhexyl) phthalate (DEHP)and polychlorinated biphenyls (PCBs) causes reproductive dysfunction in adult male mouse offspring. Reprod Toxicol 65:123–132

Finkel T (2011) Signal transduction by reactive oxygen species. J Cell Biol 194(1):7–15

Fronczak CM, Kim ED, Barqawi AB (2012) The insults of illicit drug use on male fertility. J Androl 33:515–528

Furukawa S, Fujita T, Shimabukuro M, Iwaki M, Yamada Y, Nakajima Y, Nakayama O, Makishima M, Matsuda M, Shimomura I (2004) Increased oxidative stress in obesity and its impact on metabolic syndrome. J Clin Invest 114(12):1752

Gabrielsen JS, Tanrikut C (2016) Chronic exposures and male fertility: the impacts of environment, diet, and drug use on spermatogenesis. Andrology 4(4):648–661

Gharagozloo P, Aitken RJ (2011) The role of sperm oxidative stress in male infertility and the significance of oral antioxidant therapy. Hum Reprod 26:1628–1640

Golalipour M, Azarhoush R, Ghafari S, Gharravi A, Fazeli S, Davarian A (2007) Formaldehyde exposure induces histopathological and morphometric changes in the rat testis. Folia Morphol 66:167–171

Guerin P, El Mouatassim S, Menezo Y (2001) Oxidative stress and protection against reactive oxygen species in the pre-implantation embryo and its surroundings. Hum Reprod Update 7(2):175–189

Gules O, Eren U (2010) The effect of xylene and formaldehyde inhalation on testicular tissue in rats. Asian-Aust J Anim Sci 23:1412–1420

Gunter TE, Gavi CE, Aschner M, Gunter KK (2006) Speciation of manganese in cells and mitochondria: a search for the proximal cause of manganese neurotoxicity. Neurotoxicol 27(5):765–776

Guthauser B, Boitrelle F, Plat A, Thiercelin N, Vialard F (2013) Chronic excessive alcohol consumption and male fertility: a case report on reversible azoospermia and a literature review. Alcohol Alcohol 49:42–44

Hagan S, Khurana N, Chandra S, Abdel-Mageed AB, Mondal D, Hellstrom WJ, Sikka SC (2015) Differential expression of novel biomarkers (TLR-2, TLR-4, COX-2, and Nrf-2) of inflammation and oxidative stress in semen of leukocytospermia patients. Andrology 3:848–855

Hammoud AO, Gibson M, Peterson CM, Meikle AW, Carrell DT (2008) Impact of male obesity on infertility: a critical review of the current literature. Fertil Steril 90:897–904

Harclerode J (1984) Endocrine effects of marijuana in the male: preclinical studies. NIDA Res Monogr 44:46–64

Hauser R, Chen Z, Pothier L, Ryan L, Altshul L (2003) The relationship between human semen parameters and environmental exposure to polychlorinated biphenyls and p, p′-DDE. Environ Health Perspect 111:1505–1511

Henkel RR (2011) Leukocytes and oxidative stress: dilemma for sperm function and male fertility. Asian J Androl 13:43–52

Holland MK, White IG (1988) Heavy metals and human spermatozoa, 111: the toxicity of copper ions for spermatozoa. Contraception 38:685–695

Jana K, Sen PC (2012) Environmental toxicants induced male reproductive disorders: identification and mechanism of action. In: Prof. Acree B (ed) Toxicity and drug testing. InTech. ISBN: 978-953-51-0004-1. Available from: http://www.intechopen.com/books/toxicity-and-drug-testing/environmentaltoxicants-induced-male-reproductive-disorders-identification-and-mechanism-of-action-

Jensen TK, Andersson AM, Jorgensen N, Andersen AG, Carlsen E, Petersen JH, Skakkebaek NE (2004) Body mass index in relation to semen quality and reproductive hormones among 1,558 Danish men. Fertil Steril 82:863–870

Jensen CFS, Ostergren P, Dupree JM, Ohl DA, Sonksen J, Fode M (2017) Varicocele and male infertility. Nat Rev Urol 14(9):523–533

Karila T, Hovatta O, Seppala T (2004) Concomitant abuse of anabolic androgenic steroids and human chorionic gonadotrophin impairs spermatogenesis in power athletes. Int J Sports Med 25:257–263

Knuth UA, Maniera H, Nieschlag E (1989) Anabolic steroids and semen parameters in bodybuilders. Acta Endocrinol 120:S121–S122

Kumar N, Singh AK (2015) Trends of male factor infertility, an important cause of infertility: a review of literature. J Hum Reprod Sci 8:191–196

Kunzle R, Mueller MD, Hanggi W, Birkhauser MH, Drescher H, Bersinger NA (2003) Semen quality of male smokers and nonsmokers in infertile couples. Fertil Steril 79:287–291

La Maestra S, De Flora S, Micale RT (2015) Effect of cigarette smoke on DNA damage, oxidative stress, and morphological alterations in mouse testis and spermatozoa. Int J Hyg Environ Health 218:117–122

La Vignera S, Condorelli RA, Balercia G, Vicari E, Calogero AE (2013) Does alcohol have any effect on male reproductive function? A review of literature. Asian J Androl 15:221–225

Lamb EJ, Bennett S (1994) Epidemiologic studies of male factors in infertility. Ann N Y Acad Sci 709:165–178

Lancranjan I, Popescu HI, Gavanescu O, Klepsch I, Serbanescu M (1975) Reproductive ability of workmen occupationally exposed to lead. Arch Environ Health: Int J 30:396–401

Lavranos G, Balla M, Tzortzopoulou A, Syriou V, Angelopoulou R (2012) Investigating ROS sources in male infertility: a common end for numerous pathways. Reprod Toxicol 34:298–307

Li Y, Wu J, Zhou W, Gao E (2012) Effects of manganese on routine semen quality parameters: results from a population-based study in China. BMC Public Health 12:919

Liu H, Niu R, Wang J, He Y, Wang J, China S (2008) Changes caused by fluoride and lead in energy metabolic enzyme activities in the reproductive system of male offspring rats. Res Rep Fluoride 41(3):184–191

Magnusdottir EV, Thorsteinsson T, Thorsteinsdottir S, Heimisdottir M, Olafsdottir K (2005) Persistent organochlorines, sedentary occupation, obesity and human male subfertility. Hum Reprod 20:208–215

Majzoub A, Arafa M, Elbardisi H, Al Said S, Agarwal A, Al Rumaihi K (2017) Oxidation-reduction potential and sperm DNA fragmentation levels in sperm morphologic anomalies. In: Poster presented at: 33rd annual European society for health and reproduction conference, July 3–5, Geneva, Switzerland

Manguez-Alarcon L, Hauser R, Gaskins AJ (2016) Effects of bisphenol-A on male and couple reproductive health: a review. Fertil Steril 106:864–870

Marnett L (1999) Lipid peroxidation—DNA damage by malondialdehyde. Mutat Res 424:83–95

Martini AC, Molina RIS, Estofan D, Senestrari D, de Cuneo MF, Ruiz RND (2004) Effects of alcohol and cigarette consumption on human seminal quality. Fertil Steril 82:374–377

Meeker JD, Rossano MG, Protas B, Diamond MP, Puscheck E, Daly D, Paneth N, Wirth JJ (2008) Cadmium, lead, and other metals in relation to semen quality: human evidence for molybdenum as a male reproductive toxicant. Environ Health Perspect 116:1473–1479

Meeker JD, Yang T, Ye X, Calafat AM, Hauser R (2011) Urinary concentrations of parabens and serum hormone levels, semen quality parameters, and sperm DNA damage. Environ Health Perspect 119(2):252–257

Mocarelli P, Gerthoux PM, Patterson DG Jr, Milani S, Limonta G, Bertona M, Signorini S, Tramacere P, Colombo L, Crespi C, Brambilla P, Sarto C, Carreri V, Sampson EJ, Turner WE, Needham LL (2008) Dioxin exposure, from infancy through puberty, produces endocrine disruption and affects human semen quality. Environ Health Perspect 116(1):70–77

Moustafa MH, Sharma R, Thornton J, Mascha E, Abdel-Hafez MA, Thomas AJ Jr, Agarwal A (2004) Relationship between ROS production, apoptosis and DNA denaturation in spermatozoa from patients examined for infertility. Hum Reprod 19:129–138

Muller C, Lee T, Montano M (2013) Improved chemiluminescence assay for measuring antioxidant capacity of seminal plasma. Methods Mol Biol 927:363–376

Multigner L, Catala M, Cordier S, Delaforge M, Fenaux P, Garnier R, Rico-Lattes I, Vasseur P (2005) The INSERM expert review on glycol ethers: findings and recommendations. Toxicol Lett 156:29–37

Multigner L, Brik EB, Arnaud I, Haguenoer JM, Jouannet P, Auger J, Eustache F (2007) Glycol ethers and semen quality: a cross-sectional study among male workers in the Paris Municipality. Occup Environ Med 64:467–473

Muthusami KR, Chinnaswamy P (2005) Effect of chronic alcoholism on male fertility hormones and semen quality. Fertil Steril 84:919–924

Naha N, Chowdhury AR (2006) Inorganic lead exposure in battery and paint factory: effect on human sperm structure and functional activity. J UOEH 28:157–171

Nakai N, Murata M, Nagahama M, Hirase T, Tanaka M, Fujikawa T, Nakao N, Nakashima K, Kawanishi S (2003) Oxidative DNA damage induced by toluene is involved in its male reproductive toxicity. Free Radical Res 37(1):69–76

Naviaux RK (2012) Oxidative shielding or oxidative stress? J Pharmacol Exp Ther 342:608–618

Nudell DM, Monoski MM, Lipshultz LI (2002) Common medications and drugs: how they affect male fertility. Urol Clin North Am 29:965–973

Oliva R (2006) Protamines and male infertility. Hum Reprod Update 12:417–435

Panner Selvam MK, Henkel R, Sharma R, Agarwal A (2018) Calibration of redox potential in sperm wash media and evaluation of oxidation-reduction potential values in various assisted reproductive technology culture media using MiOXSYS system. Andrology 6:293–300

Pizent A, Tariba B, Zivkovic T (2012) Reproductive toxicity of metals in men. Arch Ind Hyg Toxicol 63(Suppl):35–46

Polson D, Villalba N, Freeman K (2018) Optimization of a diagnostic platform for oxidation-reduction potential (ORP) measurement in human plasma. Redox Rep 23(1):125–129

Rahman MS, Kwon WS, Lee JS, Yoon SJ, Ryu BY, Pang MG (2015) Bisphenol-A affects male fertility via fertility-related proteins in spermatozoa. Sci Rep 5:9169

Rosa MD, Zarrilli S, Paesano L, Carbone U, Boggia B, Petretta M, Maisto A, Cimmino F, Puca G, Colao A (2003) Traffic pollutants affect fertility in men. Human Reprod 18:1055–1061

Roychoudhury S, Massanyi P (2008) In vitro copper inhibition of the rabbit spermatozoa motility. J Environ Sci Health Part A: Toxic Hazard Subst Environ Eng 43(6):651–656

Roychoudhury S, Massanyi P, Bulla J, Choudhury MD, Straka L, Lukac N, Formicki G, Dankova M, Bardos L (2010) In vitro copper toxicity on rabbit spermatozoa motility, morphology and cell membrane integrity. J Environ Sci Health Part A: Toxic Hazard Subst Environ Eng 45(12):1482–1491

Roychoudhury S, Nath S, Massanyi P, Stawarz R, Kacaniova M, Kolesarova A (2016a) Copper-induced changes in reproductive functions: in vivo and in vitro effects. Physiol Res 65(1):11–22

Roychoudhury S, Sharma R, Sikka S, Agarwal A (2016b) Diagnostic application of total antioxidant capacity in seminal plasma to assess oxidative stress in male factor infertility. J Assist Reprod Genet 33:627–635

Roychoudhury S, Agarwal A, Virk G, Cho CL (2017a) Potential role of green tea catechins in the management of oxidative stress-associated infertility. Reprod Biomed Online 34(5):487–498

Roychoudhury S, Dorsey C, Choudhury BP, Kar KK (2017b) Oxidation-reduction potential can help distinguish semen samples under oxidative stress. Hum Reprod 32:168

Roychoudhury S, Maldonado-Rosas I, Agarwal A, Esteves SC, Henkel R, Sharma R (2018) Human sperm handling in intracytoplasmic sperm injection processes: in vitro studies on mouse oocyte activation, embryo development competence and sperm oxidation-reduction potential. Andrologia e12943

Said TM, Ranga G, Agarwal A (2005) Relationship between semen quality and tobacco chewing in men undergoing infertility evaluation. Fertil Steril 84:649–653

Saleh RA, Agarwal A, Sharma RK, Nelson DR, Thomas AJ Jr (2002) Effect of cigarette smoking on levels of seminal oxidative stress in infertile men: a prospective study. Fertil Steril 78:491–499

Saleh RA, Agarwal A, Nada EA, El-Tonsy MH, Sharma R, Meyer A, Nelson DR, Thomas AJ Jr (2003) Negative effects of increased sperm DNA damage in relation to seminal oxidative stress in men with idiopathic and male factor infertility. Fertil Steril 79:1597–1605

Sarkar A, Ravindran G, Krishnamurthy V (2013) A brief review on the effect of cadmium toxicity: from cellular to organ level. Int J Biotech 3:17–36

Seftel A (2006) Male hypogonadism. Part II: etiology, pathophysiology, and diagnosis. Int J Impot Res 18:223–228

Semet M, Paci M, Saias-Magnan J, Metzlera-Guillemain C, Boissier R, Lejeune H, Perrin J (2017) The impact of drugs on male fertility: a review. Andrology 5:640–663

Sepaniak S, Forges T, Gerard H, Foliguet B, Bene MC, Monnier-Barbarino P (2006) The influence of cigarette smoking on human sperm quality and DNA fragmentation. Toxicology 223:54–60

Sharma R, Agarwal A (1996) Role of reactive oxygen species in male infertility. Urology 48:835–850

Sharpe RM (2010) Environmental/lifestyle effects on spermatogenesis. Philos Trans R Soc Lond B: Biol Sci 365:1697–1712

Sikka SC, Toor JS, Lucelia D, Yafi FA, Hellstrom W (2016) Measurement of oxidation-reduction potential (ORP) as a newer tool indicative of oxidative stress in infertile men with leukocytospermia. In: Poster presented at: 41st annual America society of andrology conference, Apr 2–5, New Orleans, LA, USA

Siu ER, Mruk DD, Porto CS, Cheng CY (2009) Cadmium-induced testicular injury. Toxicol Appl Pharmacol 238(3):240–249

Song B, Cai ZM, Li X, Deng LX, Zheng LK (2005) Effect of benzene on sperm DNA. Zhonghua Nan Ke Xue 11(1):53–55

Spira A, Multigner L (1998) The effect of industrial and agricultural pollution on human spermatogenesis. Hum Reprod 13:2041–2042

Sunanda P, Panda B, Dash C, Ray PK, Padhy RN, Routray P (2014) Prevalence of abnormal spermatozoa in tobacco chewing sub-fertile males. J Hum Reprod Sci 7:136–142

Toor JS, Daniels L, Yafi FA, Hellstrom W, Sikka SC (2016) Semen oxidative reduction potential (ORP) is related to abnormal semen parameters in male factor infertility. In: Poster presented at: 41st annual America society of andrology conference, Apr 2–5, New Orleans, LA, USA

Trussell JC (2013) Optimal diagnosis and medical treatment of male infertility. Semin Reprod Med 31:235–236

Turner TT, Lysiak JJ (2008) Oxidative stress: a common factor in testicular dysfunction. J Androl 29:488

Ufer C, Wang CC, Borchert A, Heydeck D, Kuhn H (2010) Redox control in mammalian embryo development. Antioxid Redox Signal 13:833–875

Villegas J, Schulz M, Soto L, Iglesias T, Miska W, Sanchez R (2005) Influence of reactive oxygen species produced by activated leukocytes at the level of apoptosis in mature human spermatozoa. Fertil Steril 83:808–810

Vine MF, Tse CK, Hu P, Truong KY (1996) Cigarette smoking and semen quality. Fertil Steril 65(4):835–842

Whan LB, West MC, McClure N, Lewis SE (2006) Effects of delta-9-tetrahydrocannabinol, the primary psychoactive cannabinoid in marijuana, on human sperm function in vitro. Fertil Steril 85:653–660

Xiao GB, Pan CB, Cai YZ, Lin H, Fu ZM (2001) Effect of benzene, toluene, xylene on the semen quality and the function of accessory gonad of exposed workers. Ind Health 39(2):206–210

Zhou DX, Qiu SD, Zhang J, Tian H, Wang HX (2006) The protective effect of vitamin E against oxidative damage caused by formaldehyde in the testes of adult rats. Asian J Androl 8:584–588

Zini A, San Gabriel M, Baazeem A (2009) Antioxidants and sperm DNA damage: a clinical perspective. J Assist Reprod Genet 26:427–432

System Network Biology Approaches in Exploring of Mechanism Behind Mutagenesis

Anukriti, Swati Uniyal, Anupam Dhasmana, Meenu Gupta, Kavindra Kumar Kesari, Qazi Mohd. Sajid Jamal and Mohtashim Lohani

Abstract Mutagenesis is the alteration of the genetic material by the help of mutagens. Mutations that are capable of inducing any diseases have a large impact on the biological systems. Whenever mutation occurs, it not only affects any particular gene or protein, but also affects the whole system related to that gene. Changes in one system will further bring out changes in the adjacent systems, which works in coordination with the mutated system. Thus, a single mutation can have an impact on more than one system. System network biology helps in providing a new perspective of inspection of these biological systems in the form of networks with the help of mathematical representations. In this chapter, we deal with different properties of the networks that help in analyzing the network-graph and finding the most probable network that best describes the process. Here we tried to investigate the candidate protein molecule that may act as a target protein with the help of network

Anukriti
Himalayan School of Biosciences, Swami Rama Himalayan University, Dehradun, India

S. Uniyal
School of Biotechnology, Gautum Buddha University, Greater Noida, India

A. Dhasmana (✉)
Himalayan School of Biosciences & Cancer Research Institute, Swami Rama Himalayan University, Dehradun, India
e-mail: anudhas007@gmail.com

M. Gupta
Department of Radiotherapy, Cancer Research Institute, Swami Rama Himalayan University, Dehradun, India

K. K. Kesari
Department of Chemical Engineering, Aalto University, Aalto, Espoo, Finland

Q. Mohd. S. Jamal
Department of Health Informatics, College of Public Health and Health Informatics, Qassim University, King Abdul Aziz Road, Al Bukayriyah, Saudi Arabia

M. Lohani
Department of Emergency Medical Services, Collage of Applied Medical Sciences, University of Jazan, Jazan, Kingdom of Saudi Arabia

© Springer Nature Switzerland AG 2019
K. K. Kesari (ed.), *Networking of Mutagens in Environmental Toxicology*, Environmental Science,
https://doi.org/10.1007/978-3-319-96511-6_6

analysis. For this, we used various datasets and software that would be used in the reconstruction of different biological networks and pathways.

1   Introduction

Mutations are the spontaneous changes that occur in nature. These alterations can be both, useful as well as harmful. These alterations take place on the basis of evolution at various biological levels. Mutations may have developed artificially through spontaneous hydrolysis. Mutations could generate ROS and DNA adducts or produce error in repair and replication process of DNA. The mammalian cells work on rescue mechanisms where it may participate to repair the mutated sites of the DNA. If the repair mechanisms fail to correct the errors in DNA, the mutated DNA gets copied in the daughter cells. Moreover, if the damage is large and irreparable, then the cells has the mechanism to induce programmed cell death.

Artificially induced mutations help in understanding the structural or functional relationships between different proteins. Polymerase chain reaction (PCR) is the most widely used technique for artificially induced mutations. Real time-PCR is very simple but robust and highly sensitive technique used for the detection of mutations. Although, for diagnosis of multifactorial diseases like cancer, molecular tests are widely used. Moreover, for molecular characterization of tumors, the key parameters used as genetic mutations and the gene expression profiles. These helps in accurate prognosis of diseases and selection of treatments (Morlan et al. 2009). Medical science has now entered into the era of systems biology. In systems biology, biological systems are represented as networks. These networks represent complex interaction between different entities. In medical sciences, network-based methods are being used for the detection of mutations and also being used for prediction of stability of mutation induced proteins (Frenz 2005). Network biology not only used for the representation but also for the analysis of interactions between the entities of various biological systems using graph theory (Stam and Reijneveld 2007).

2   Graph Theory

Graph theory is the mathematical study of graphs, which consists of vertices, also known as nodes and edges or lines. In present scenario, graphs can be used to represent almost all the practical problems ranging from social and physical to the biological and other information systems. Computational neuroscience graphs are useful method to represent various physical and functional connections between different parts of the brain. Nodes of different parts and areas of the brain are presented by the functional interactions between these nodes, which may be presented by the edges (Kang et al. 2011). With the help of structural analysis of graphs and by calculating the distances, one can be able to determine the exact expression of the entities (Gao et al. 2017).

3 What Is System Network Biology?

Systems biology is a science that mainly focuses on the building of mathematical models for cellular networks (Villaverde et al. 2013). A network comprises of nodes and edges. The nodes represent the genes or proteins involved in process while the edges represent physical interactions between them (Bosley et al. 2013). Immensely complex networks of protein-protein interactions govern all the biological processes inside the organisms. These networks are highly packed and comprises of nodes and edges. For the study of biological networks, there are some standard steps that must be followed (Lv et al. 2013).

At very first step, it is important to find out candidate genes that are linked with the biological system under study. For this, literature survey along with text mining and data mining must be performed. There are various databases, which may help in finding the associated genes like t3db for carcinogens, Reactome, KEGG etc.

Next step is to scan the protein-protein interactions. For this databases like STRING can be used. One can even construct the protein-protein interaction network by the help of STRING database (Szklarczyk et al. 2015).

Once PPI network has been constructed, the next and the most important step is topological analysis of PPI network. For topological analysis, one can find the node with this highest degree and that will be hub node; the node with maximum betweenness centrality which would depict the bottleneck protein etc. After topological analysis, one can construct the backbone network from nodes with large betweenness centralities followed by its validation.

3.1 Types of Networks

Different types of networks are created depending on the requirement.

3.1.1 Directed Network

Directed networks or graphs are those in which direction of the movement of information is known. It is also known as a digraph. By definition, a directed graph is a graph that has nodes or vertices and directed edges. In biological systems, such networks are of great importance as these help in predicting the possible drug targets.

3.1.2   Undirected Network

Undirected graphs are the graphs in which the edges have no particular orientation. In mathematical terms, an undirected graph is one, which consists of a set of N nodes and a set of E edges, which are unordered pairs of elements of N.

3.1.3   Weighted Network

These are the networks in which each edge has assigned a weight. The advantage of weighted network is that it helps in estimation of relationship between elements or nodes in network.

3.1.4   Unweighted Network

The networks in which the edges are not weighted are unweighted networks or graphs. If one has the local information, weighted networks could be created from the unweighted networks. This could be done by the distribution of the weight to the edges.

3.2   Types of Network Edges

3.2.1   Undirected Edges

This type of edge is found in protein-protein interaction networks (PPINs). Undirected edges represent the simple relationship between two nodes. These do not represent the direction of flow of information between nodes.

3.2.2   Directed Edges

This is a kind of connection found, for example, in metabolic or gene regulation networks. There has a clear flow of signal implied and the network can be organized hierarchically.

3.2.3   Weighted Edges

Weighted edges represent the qualitative relation between the two nodes. Directed as well as undirected edges can have weights. Weights depict various information like how reliable the interaction is, how much sequence similarity presents between the two genes etc. Edges can also be weighted by their centrality values or several other topological parameters.

4 Properties/Topological Analysis of Biological Networks

To understand a complex system network, analysis could be done in a bottom-up approach in systems biology. Analysis starts from individual components and reaches to their connections. Interaction of each element with other elements has been done in network analysis. Network analysis could be done locally as well as globally. Local topological measurements of networks include degree, shortest path length and size used to analyze the network. Whereas for the global network analysis, terms like mean shortest path length and average degree may be used. The mean shortest path length depicts the average count of steps required or used to join each and every pair of nodes present in network by using their shortest path. The average degree of network is an average of degrees of all nodes that would present in the network (Ran et al. 2013).

4.1 Size

Size of network has the total number of nodes that may present in that network.

4.2 Degree

Degree defines the connectivity of individual nodes in network with other nodes. It is the number of connections of a node with other nodes in a network. If the network is directed network, then each node has two types of degrees, namely, in-degree and out-degree presented in Fig. 1. In-degree is the number of edges that are coming to node (Fig. 1a) while out-degree is defined as the number of edges having that node as a source (Fig. 1b).

4.3 Shortest Path

Shortest path is also known as the geodesic path. It is the path in a network that connects two nodes with minimum number of edges required.

Fig. 1 Shows directed
network with different nodes
namely **a** in-degree is the
number of edges that are
coming to the node,
b out-degree is defined as the
number of edges having that
node as a source

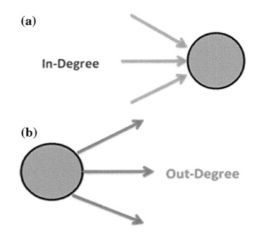

4.4 Scale-Free Network

Scale-free networks are the networks in which connections of each node with other nodes follow the power law. Scale free networks tend to be more robust. Even if any node has been removed, the network would not be fallen apart and loses its connectivity, but this only hold true when the nodes removed may not the major hub nodes.

4.5 Transitivity

Transitivity is rarely found in real networks. It refers to connections between all the nodes and the edges by which they may connected. To explain this in simple words, if nodes A and B are connected with an edge and node B is also connected with another node, say C, then node A and node C are also supposed to be connected. This is rarely found in real life network problems. Those graphs, which are intransitive, leads to the formation of clusters in the graph. The measure of transitivity or the transitivity index is also known as the clustering index (Frank 2010). This index ranges from 0 to 1 where 1 represents a complete transitive graph.

4.6 Centralities

Centrality is one of the most important properties in a graph. It helps in identification of the most important nodes and edges within the graph. Centrality analysis in biological networks is important as it gives an estimate of which proteins are important in a PPIN. There are various types of centralities, which are analyzed depending on the type of analysis. The algorithm for the calculation also changes depending on the type of graph under study (Kalna and Higham 2007). For weighted and unweighted graphs, separate algorithms are followed. There are two most important centralities, which are mostly estimated, in a biological network. These are: betweenness centrality and closeness centrality.

4.6.1 Betweenness Centrality

It is the measure of times; a node acts as a bridge in the shortest path between two other nodes (Dietz et al. 2010). The nodes with high betweenness centrality control the flow of information between two other nodes. Such nodes play important roles in biological pathways and hence can be the most potent targets for drug discoveries. There is no standard formula for the calculation of betweenness centrality. The most basic way to find betweenness centrality is by calculating the number of shortest paths that pass through that node in graph and then dividing it by total number of shortest paths (Borgattia and Everett 2006). Betweenness centrality can further be divided into node betweenness centrality and edge betweenness centrality. Node betweenness centrality tells the most central node in network while edge betweenness centrality tells the fraction of path through which maximum information passes between two nodes in whole network (Brandes 2008; Dietz et al. 2010).

4.6.2 Closeness Centrality

This defines the closeness of node that how close a node is with other nodes in a graph (Borgatti 2005). It could be also defined the closeness centrality as how fast we can pass a piece of information from given node to another. It is an inverse to the farness.

4.7 Bottleneck Proteins

In network, those proteins or nodes, which have high betweenness centrality, are known as bottleneck proteins. Bottleneck proteins are of huge importance in the topological analysis of the biological networks (Yu et al. 2007; Zhu et al. 1999). These proteins play significant role of connectors between different proteins and these can be considered as dynamic components in any interaction networks (Zhu et al. 1999).

4.8   Clustering Coefficient

Cluster analysis is done to create modules that have mutually exclusive properties from each other (Gavin et al. 2006; Bullmore and Sporns 2009). It has important applications to club different elements with similar properties. For example, if protein-protein interaction network for the cell cycle regulatory proteins has been created and after performing cluster analysis, it will most likely to get the clusters or modules that are involved in different pathways like the apoptotic pathway, proteins involved in pathways of cancer, NF-KB signaling pathway etc.

Clustering coefficient can be defined as measures of degree in a graph to which nodes tend to be in a cluster (Barabási and Oltvai 2004; Opsahl and Panzarasa 2009). There are two types of clustering coefficient that are used for network analysis namely local clustering coefficient and global clustering coefficient. Local and global clustering coefficients are calculated differently for directed and undirected networks (Kalna and Higham 2007; Krot and Prokhorenkova 2015; Prokhorenkova and Samosvat 2014; Pržulj et al. 2004).

4.9   Hub Proteins

Hub proteins are those proteins that have maximum degree and the nodes that are highly connected with other nodes in network (Vallabhajosyula et al. 2009; He and Zhang 2006; Zotenko et al. 2008). There are two types of hub proteins in any network namely date hub and party hub. There is a clear distinction between the party hub and the date hub. Party hubs are also known as static hubs where as another name of date hub is dynamic hub (Vandereyken et al. 2018; Frank 2010; Chang et al. 2013). Party hubs interact with all their partners at the same time where as date hubs interact at different times or different locations with their partners. Date hubs act like connectors that organize different autonomous modules while party hubs appear as the centre for each autonomous module (Dietz et al. 2010; Goel and Wilkins 2012).

4.10   Clique

It is a subset of an undirected graph. In protein-protein interaction network, Spirin and Mirny in 2003, detected tightly linked protein clusters that had few interactions outside the cluster (Spirin and Mirny 2003). As clique is the subset of the undirected graph, it is important to find the maximum clique in a graph for proper analysis of a network. To find the maximum number of clique, systematic inspection of all the subsets needs to be done. By this way of finding maximum number of clique represents Brute-force search but this might be very time consuming (Wu 2013).

5 Case Study

Human beings are surrounded by numerous chemicals and due to these chemicals contaminations serious mutations may occurs. Agencies like EPA have reported that diesel exhausts contain fine particles that pose serious threats to human health. As per the reports from National Toxicology Program, there are around 200–300 carcinogens that include pesticides, cigarette smoke, PAH etc. Nicotine is highly addictive compound present in cigarette smoke (Buisson and Bertrand 2002). Exposures to nicotine may cause DNA damage (Hecht 2003; Howard et al. 1998; Pfeifer et al. 2002; Zhu et al. 1999) as well as changes in the cell cycle regulatory machinery. By the help of databases like PUBMED and t3db (Wishart et al. 2015; Lim et al. 2010; Lu 2011) in this article we were able to find around 300 genes and they were, getting up-regulated and down regulated (Table 3) because of the interaction of nicotine. With the help of STRING, we have created a topological network (Fig. 2) and also performed further analysis to find the hub proteins that could be potent target candidates for nicotine. STRING software not only allows the user to create topological network but it also performs enrichment analysis of the query proteins. STRING obtains these results from the Gene Ontology, KEGG pathways, InterPro and Pfam domains based on the enriched P values. Topological analysis was done with the help of Cytoscape (stand-alone software) (Shannon et al. 2003). We have created the clusters by using MCODE App of the cytoscape software. MCODE creates clusters based on the topology. It helps in finding densely inter-connected regions. Figures 3 and 4 represent top 2 clusters of nicotine network that were generated by MCODE. The clusters were ranked based on the cluster's computed scores and numbered according to their ranks. The cluster 1 (Fig. 3) was highest ranked cluster with computed score of 38. It had 38 nodes and 703 edges. Cluster 2 (Fig. 4) was second ranked cluster with score of 37.263, 39 nodes and 709 edges. We have also generated the clusters based on KEGG enrichment pathways (Table 1), which has been obtained from STRING database. Table 1 consists of only those enriched KEGG pathways for which the corrected P value lies above PPI-enriched P value. Figure 5 represents cluster network generated by the genes that were hampered by nicotine action and they were related to Parkinson's disease. The enriched network pathway for the Parkinson's disease consists 44 nodes, 669 edges. The average node degree was 30.4 and the average clustering coefficient was 0.977. Figure 6 represents the cluster network generated by genes that were associated with Alzheimer's disease. The enriched network pathway for Alzheimer's disease consisted 42 nodes, 501 edges, and had average node degree of 23.9 and an average local clustering coefficient of 0.857. Figure 7 represents cluster network generated by genes that were associated with the Huntington's disease. The enriched network pathway for the Huntington's disease consisted 39 nodes, 520 edges. The average node degree was 26.7 and the average local clustering coefficient was 0.951. Figure 8 shows the genes that got hampered by the action of nicotine and they causes cancer. This network cluster consisted 29 nodes, 53 edges, and an

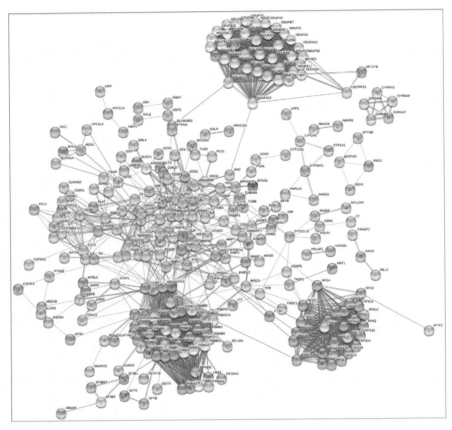

Fig. 2 Complete protein topology network of nicotine. Total no. of nodes present in this network are 422, edges are 2260, avg. node degree are 10.7 and avg. local clustering coefficient is 0.528

average node degree of 2.86 and an average local clustering coefficient of 0.336. We have also tried to analyze important nodes in the main network with the help of cyto-Hubba App of cytoscape software. CytoHubba helps to predict an important node in network and its subnetworks using various algorithms and displays their results in table format. It calculates the degree, centralities, bottleneck, clustering coefficient etc. of all nodes involved in selected network. Table 2 represents a list of important nodes of nicotine topological network along with their properties.

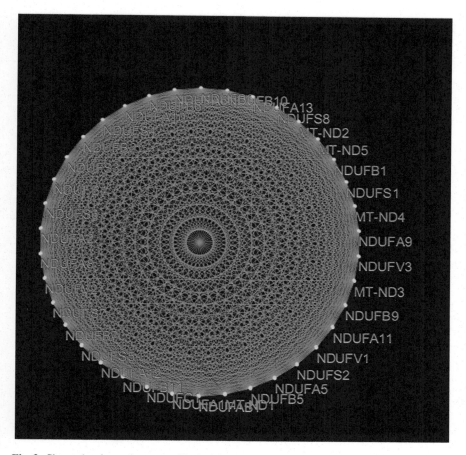

Fig. 3 Shows the cluster 1 generated by MCODE

Table 1 Enriched-KEGG pathways

Pathway ID	Pathway description	Observed gene count	P value
5012	Parkinson s disease	44	1.24E−37
3050	Proteasome	29	4.34E−37
4932	Non-alcoholic fatty liver disease (NAFLD)	42	3.22E−34
190	Oxidative phosphorylation	39	9.74E−33
5010	Alzheimer s disease	42	2.94E−32
5016	Huntington s disease	39	7.11E−27
5200	Pathways in cancer	37	7.58E−16

Table 2 CytoHubba analysis of important nodes

Name	Degree	Betweenness centrality	Bottleneck	Clustering coefficient	Closeness centrality
TP53	49	12930.9559	49	0.3767	125.58333
NFKB1	43	6202.05525	27	0.55925	121.99286
PSMB8	41	1188.83354	10	0.82073	114.40952
PSMD14	40	4685.81213	7	0.86154	116.99286
NDUFA12	39	11763.02235	39	0.90553	91.42024
PSMA3	39	1458.85911	1	0.90553	115.43333
NDUFA9	39	374.45427	2	0.90688	82.68373
PSMD6	39	248.55397	4	0.90688	112.49286
PSMA4	39	215.88633	2	0.90688	112.15952
PSMB4	39	215.88633	1	0.90688	112.15952
PSMC1	39	133.57151	2	0.90688	112.15952
NDUFV1	38	5646.88867	26	0.94737	87.91111
CCND1	38	3655.97494	12	0.62731	116.70952
UBE2C	38	2528.14841	8	0.67425	111.91667
PSMB3	38	57.38508	2	0.95306	111.65952
PSMD2	38	57.38508	1	0.95306	111.65952
PSMB7	38	57.38508	1	0.95306	111.65952
PSMA1	38	57.38508	1	0.95306	111.65952
PSMA7	38	57.38508	1	0.95306	111.65952
PSMC5	38	57.38508	1	0.95306	111.65952
PSMA5	38	57.38508	1	0.95306	111.65952
PSMC4	38	57.38508	1	0.95306	111.65952
PSMD8	38	57.38508	1	0.95306	111.65952
PSMD4	38	57.38508	1	0.95306	111.65952
PSMB6	38	57.38508	1	0.95306	111.65952
PSMD1	38	57.38508	1	0.95306	111.65952
PSMB2	38	57.38508	1	0.95306	111.65952
PSMB1	38	57.38508	1	0.95306	111.65952
PSMB5	38	57.38508	1	0.95306	111.65952
PSMC6	38	57.38508	1	0.95306	111.65952
PSMA6	38	57.38508	1	0.95306	111.65952

Table 3 List of up-regulated and down-regulated genes

Up-regulated genes						Down-regulated genes		
OAS1	CHD2	GPR50	MMP7	PRPF38B	C3AR1	APEH	CHRNA6	PTPRA
HMGCR	F2	GPR56	MMP9	PPP2R3A	TSHR	APP	CFAP43	RAB6A
ARC	CCDC159	GSDMA	MMP13	SPTLC3	FTH1P20	AMHR2	C5AR1	RFXAP
ADAM19	CCDC167	GATA2	TMEM41B	PRKAR2B	TWIST1	BTF3	CNTN1	RBBP6
AMD1	C3AR1	GLIPR1	MT1E	PKN2	C7	BAD	CKMT2	RARG
ADRB1	C7	GPNMB	MTIF2	EIF2AK2	UGP2	BCL2A1	CYP11A1	ARHGEF7
ADRB2	CLMP	CGA	NDUFA5	PROX1	UPP1	BRCA1	CYP26A1	RPL7A
ATRX	CCND1	GM2A	NFKB1	S100A14	VCAM1	BTK	DDB1	RUNX1T1
ANGPT4	CCNG2	GOLGA4	NOS2	PTGS2	EZR	CDH1	DEDD	S100A10
ACE	CDK13	GRAMD1B	NOS3	GRAMD1B	VWF	CREB1	DEFB129	SERPINH1
AATK	CDK6	GDF1	NTSR1	TAOK1	WNT7B	CARTPT	DHPS	STAT2
ALOX5	CRCT1	GCH1	OAS1	PTBP2	XIST	CASP9	DNASE2	SIVA1
ARGLU1	CYP1A1	GNG10	ORC2	RBM25	YES1	CD4	DTYMK	SLC19A1
BIRC5	CYP2A6	HDDC2	SLC22A18AS	RAC2	ZNF146	CRABP2	DHFR	SLC25A48
BMI1	CYP4F3	HSP90AA1	SERPINE1	CCND1	PAX8	CCL16	DVL2	SLC9A3R2
BTBD11	DCAF16	HEY1	HP1BP3	RGS1	CXCR4	CCR6	DFFB	STAR
CALR	DDX17	HP1BP3	HDDC2	RLF	MLPH	CXCL2	EML4	SREBF1
CAPZA2	DHRS3	HPCAL4	NIN	RPS25	PANK3	CXCR2	EDNRA	STIP1
CASP14	DDIT4	HINT1	PHF21A	S100A2	CLMP	CLIC1	EPHA3	SUPT4H1
CASP3	DBH	HOXB9	ANGPT4	MSMO1	NANOG	CHRNA3	EPHB3	TFPI2
CAMP	DYNC1LI2	CXCL8	HPCAL4	SCD	MED28	FEZ1	KMT2C	TFDP1
CTSH	ENTPD3	INSIG1	PDGFB	CCL5	KDM7A	FN1	MST1	TAGLN2
CBL	EVPLL	JUNB	UFM1	TINAGL1	CALR	FLT3	MADD	TUBA4A

(continued)

Table 3 (continued)

Up-regulated genes						Down-regulated genes		
OAS1	CHD2	GPR50	MMP7	PRPF38B	C3AR1	APEH	CHRNA6	PTPRA
CEBPB	EFNA2	MALAT1	SF3B6	EVPLL	SLC7A5	GPX1	MYL6	TUBB
CD46	EMP1	KRT6B	PIGF	BMI1	CAMP	GUCA1A	NPAS3	TNFRSF1A
CD59	EPO	KRT15	PIK3CA	SLCO2A1	CAPZA2	HLTF	NME3	TP53
CKS2	EIF2AK2	TRNP1	PLAT	SNAI2	CASP3	HMGA1	NFKBIE	UQCRFS1
CGREF1	EZR	KRT85	PLAU	SNAI1	ZCCHC9	IGHM	NEAT1	UBE3A
CENPE	FAM129A	LAMB1	PLAUR	SNAPC3	OGT	ITGA2B	NFYB	UBE2C
CCL5	FAM83F	LDLR	PLD2	SNCA	STC2	ITGAL	OLFM4	UBE2I
CCR5	FAS	LGALS1	PNMT	SOS1	CDK13	ITGB3	OPRK1	GALE
CXCL8	FASLG	LIG4	TREM1	SPRR2D	CBL	IFNAR1	PARK2	UNC119
CXCR3	FAT1	ST20	PON2	STAC	ADAM19	IFI6	PNPLA4	MYBL2
CXCR4	FAT2	CYP4F3	POU2F1	STAT5B	WISP2	IL10	PTX3	VRK2
CLCC1	FOS	SMAD9	DDIT4	STAT6	CCNG2	IL3RA	PLD1	VIM
CLCN3	FOSB	CD46	CRCT1	STXBP3	ZNF468	IL9	PTAFR	ZFR
CHRNA5	FTH1P20	MME	ATRX	ZEB1	SPAG9	KLF11	POU2AF1	
CHRNA7	FLRT2	MMP1	POU5F1	TFRC	SH2D2A	LTF	PRSS1	
CHRNB2	FOLR1	MMP2	DCAF16	TH	SCAF11	LITAF	PSMB6	
CBX3	FAH	MMP3	ARGLU1	TPR	LRRFIP1	LTB	PRKCSH	

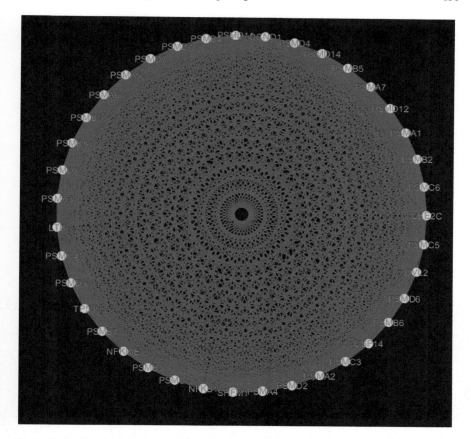

Fig. 4 Shows the cluster 2 generated by MCODE

6 Statistical Analysis of Primary Network

The network consisted 322 nodes and 2273 edges with an average node degree of 10.8 and an average clustering coefficient of 0.514. The network diameter was 9 (Fig. 2). The TP53 had highest degree, 49; bottleneck, 49; betweenness centrality, 12930.9559; closeness centrality, 125.58333; and clustering coefficient, 0.3767 (Table 2). Data based on cytoscape, we can predict that TP53 is hub protein with highest degree. Since their cluster coefficient was less than 0.5 therefore, it proposes date hub protein. It was also the bottleneck protein with the largest betweenness centrality of 12930.9559.

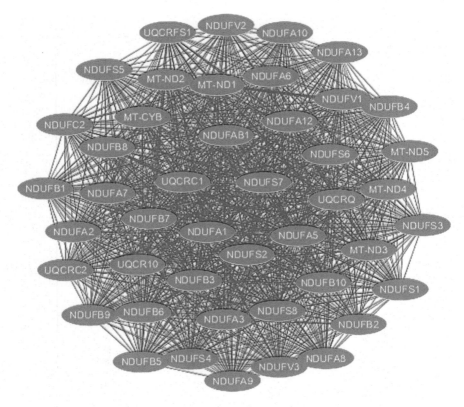

Fig. 5 Network of the genes associated with Parkinson's disease

7 Conclusions and Future Aspects

Today we are surrounded by an environment that contains enormous types of mutation causing chemicals. There are many sources such as chimney exhaust of industries, pesticides, insecticides used in the agricultural fields, diesel exhausts of vehicles, cigarette smoke, which may release hazardous chemicals in the environment. Although these are potent of causing serious damages to the human health. They are potent enough to cause serious DNA damages and lethal changes in the cell cycle that are not possible to repair or revert and leads to deadly diseases like cancer. Different chemicals have different mechanisms or mode of actions in our human body. The high-throughput technologies like micro arrays, protein chips or yeast two-hybrid screening have provided us enormous quantitative data revealing the basic design and structure of living systems. This knowledge has opened a new era of network biology, where it is becoming possible for the scientists to predict most probable pathways of the disease expression (disease network). With system biology, the pre-

Fig. 6 Network of the genes associated with Alzheimer's disease

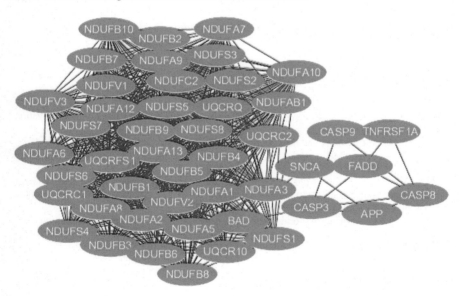

Fig. 7 Network of the genes associated with Huntington's disease

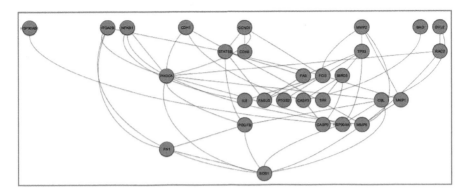

Fig. 8 Network of the genes associated with cancer

diction of most probable targets for the drugs leading to the development of targeted drug delivery methods.

Systems network biology has opened new avenues for the development of highly accurate and personalized medicine and drug development by the help of perturbation-based modeling. Perturbation in concentration of single entity in biological system could affect the physical and functional interactions of other components and thus will ultimately affect the whole interactome.

References

Barabási LS, Oltvai ZN (2004) Network biology: understanding the cell's functional organization. Nat Rev Genet 5:101–113

Borgatti SP (2005) Centrality and network flow. Soc Netw 27:55–71

Borgattia SP, Everett MG (2006) A graph-theoretic perspective on centrality. Soc Netw 28(4):466–484

Bosley AD, Das S, Andresson T (2013) A role for protein–protein interaction networks in the identification and characterization of potential biomarkers (Chap. 21). In: Proteomic and metabolomic approaches to biomarker discovery, pp 333–347

Brandes U (2008) On variants of shortest-path betweenness centrality and their generic computation. Soc Netw 2:136–145

Buisson B, Bertrand D (2002) Nicotine addiction: the possible role of functional upregulation. Trends Pharmacol Sci 23(3):130–136

Bullmore DE, Sporns O (2009) Complex brain networks: graph theoretical analysis of structural and functional systems. Nat Rev Neurosci 10:186–198

Chang X, Xu T, Li Y, Wang K (2013) Dynamic modular architecture of protein-protein interaction networks beyond the dichotomy of 'date' and 'party' hubs. Sci Rep 3:1691

Dietz KJ, Jacquot JP, Harris G (2010) Hubs and bottlenecks in plant molecular signalling networks. New Phytol 188(4):919–936

Frank O (2010) Transitivity in stochastic graphs and digraphs. J Math Soc 7(2):199–213

Frenz CM (2005) Neural network-based prediction of mutation-induced protein stability changes in Staphylococcal nuclease at 20 residue positions. Proteins Struct Funct Bioinform 59(2):147–151

Gao W, Wu H, Siddiqui MK, Baig AQ (2017) Study of biological networks using graph theory 1–8 (in press). https://doi.org/10.1016/j.sjbs.2017.11.022

Gavin AC et al (2006) Proteome survey reveals modularity of the yeast cell machinery. Nature 440:631–636

Goel A, Wilkins MR (2012) Dynamic hubs show competitive and static hubs non-competitive regulation of their interaction partners. PLoS ONE 7(10):e48209

He X, Zhang J (2006) Why do hubs tend to be essential in protein networks? PLoS Genet 2(6):e88

Hecht SS (2003) Tobacco carcinogens, their biomarkers and tobacco-induced cancer. Nat Rev Cancer 3(10):733–744

Howard DJ, Briggs LA, Pritsos CA (1998) Oxidative DNA damage in mouse heart, liver, and lung tissue due to acute side-stream tobacco smoke exposure. Arch Biochem Biophys 352(2):293–297

Kalna G, Higham DJ (2007) A clustering coefficient for weighted networks, with application to gene expression data. AI Commun—Netw Anal Nat Sci Eng 20(4):263–271

Kang U, Papadimitriou S, Sun J, Tong H (2011) Centralities in large networks: algorithms and observations. In: Proceedings of the 2011 SIAM international conference on data mining, pp 119–130

Krot A, Prokhorenkova LO (2015) Local clustering coefficient in generalized preferential attachment models. In: Gleich DF, Komjathy J (eds) Algorithms and models for the web graph. Springer International Publishing Switzerland, pp 15–28. https://doi.org/10.1007/978-3-319-26784-5_2

Lim E, Pon A, Djoumbou Y, Knox Craig, Shrivastava S, Guo AC, Neveu V, Wishart DS (2010) T3DB: a comprehensively annotated database of common toxins and their targets. Nucl Acids Res 38:D781–D786

Lu Z (2011) PubMed and beyond: a survey of web tools for searching biomedical literature. Database 2011(1):baq036

Lv YW, Jing Wang J, Sun L, Zhang JM, Cao L, Ding YY, Chen Y, Dou JJ, Huang J, Tang YF, Wu WT, Cui WR, Lv HT (2013) Understanding the pathogenesis of kawasaki disease by network and pathway analysis. Comput Math Methods Med 2013:1–17

Morlan J, Baker J, Sinicropi D (2009) Mutation detection by RT-PCR: a simple, robust and highly selective method. PLoS ONE 4(2):e4584

Opsahl T, Panzarasa P (2009) Clustering in weighted networks. Soc Netw 31(2):155–163

Pfeifer GP, Denissenko MF, Olivier M, Tretyakova N, Hecht SS, Hainaut P (2002) Tobacco smoke carcinogens, DNA damage and p53 mutations in smoking-associated cancers. Oncogene 21(48):7435–7451

Prokhorenkova LO, Samosvat E (2014) Global clustering coefficient in scale-free networks, pp 1–2. https://doi.org/10.1007/978-3-319-13123-8_5

Pržulj N, Wigle DA, Jurisica I (2004) Functional topology in a network of protein interactions. Bioinformatics 20(3):340–348

Ran J, Li H, Fu J, Liu L, Xing Y, Li X, Shen H, Chen Y, Jiang X, Li Y, Li H (2013) Construction and analysis of the protein-protein interaction network related to essential hypertension. BMC Syst Biol 7:32

Shannon P, Markiel A, Ozier O, Baliga NS, Wang JT, Ramage D, Amin N, Schwikowski B, Ideker T (2003) Cytoscape: a software environment for integrated models of biomolecular interaction networks. Genome Res 13(11):2498–2504

Spirin V, Mirny LA (2003) Protein complexes and functional modules in molecular networks. Proc Natl Acad Sci U S A 100(21):12123–12128

Stam CJ, Reijneveld JC (2007) Graph theoretical analysis of complex networks in the brain. Nonlinear Biomed Phys 1(3):1–19

Szklarczyk D, Franceschini A, Wyder S, Forslund K, Heller D, Huerta-Cepas J, Simonovic M, Roth A, Santos A, Tsafou KP, Kuhn M, Bork P, Jensen LJ, von Mering C (2015) STRING v10: protein-protein interaction networks, integrated over the tree of life. Nucl Acids Res 43:D447–D452

Vallabhajosyula RR, Chakravarti D, Lutfeali S, Ray A, Raval A (2009) Identifying hubs in protein interaction networks. PLoS ONE 4(4):e5344

Vandereyken K, Leene JV, Coninck BD, Cammue BPA (2018) Hub protein controversy: taking a closer look at plant stress response hubs. Front Plant Sci 9:694

Villaverde AF, Ross J, Banga JR (2013) Reverse engineering cellular networks with information theoretic methods. Cells 2(2):306–329

Wishart D, Arndt D, Pon A, Sajed T, Guo AC, Djoumbou Y, Knox C, Wilson M, Liang Y, Liu JGY, Goldansaz SA, Rappaport SM (2015) T3DB: the toxic exposome database. Nucl Acids Res 43(D1):D928–D934

Wu Q (2013) The maximum clique problems with applications to graph coloring. Artificial Intelligence [cs.AI]. Université d'Angers. English

Yu H, Kim PM, Sprecher E, Trifonov V, Gerstein M (2007) The importance of bottlenecks in protein networks: correlation with gene essentiality and expression dynamics. PLoS Comput Biol 3(4):e59

Zhu CQ, Lam TH, Jiang CQ, Wei BX, Lou X, Liu WW, Lao XQ, Chen YH (1999) Lymphocyte DNA damage in cigarette factory workers measured by the Comet assay. Mutat Res/Genet Toxicol Environ Mutagen 444(1):1–6

Zotenko E, Mestre J, O'Leary DP, Przytycka TM (2008) Why do hubs in the yeast protein interaction network tend to be essential: reexamining the connection between the network topology and essentiality. PLoS Comput Biol 4(8):e1000140

Ecotoxicological Effects of Heavy Metal Pollution on Economically Important Terrestrial Insects

Oksana Skaldina and Jouni Sorvari

Abstract Pollution is among the major anthropogenically induced drivers of environmental change. Heavy metals, released from industry and transport, can contaminate aquatic and terrestrial environments, inducing further ecotoxicological effects in different organisms. Insects play crucial ecological roles in maintenance of ecosystem structure and functioning and deliver such ecosystem services as food provisioning, plant pollination, dung burial, pest control and wildlife nutrition. Economically important terrestrial insects vary in an ability to accumulate heavy metals and demonstrate substantial difference in heavy metal tolerance. Despite global pollinator decline, only limited information is available about effects of heavy metals on wild bees. Ants, wasps and beetles are key-predatory insect groups in many terrestrial ecosystems. Responses in ants are investigated to higher extent than in wasps and revealed ecotoxicological effects of heavy metal pollution in beetles are biased to model species. Insect pests such as aphids and butterfly larvae respond to heavy metal pollution with modifications in their morphology and physiology, however more studies are needed to understand general directions of adaptations in this functional group of economically important insects. When investigated the problem of insect decline, heavy metal pollution should be thoroughly considered. In addition to natural habitat transformation, use of insecticides and modifications in agriculture, ecotoxicological effects of heavy metals on useful insects might have direct consequences to food security, agricultural economy and human welfare.

Keywords Ecosystem services · Ecotoxicology · Insecta · Heavy metals · Pests · Pollinators · Predators

O. Skaldina (✉) · J. Sorvari
Department of Environmental and Biological Sciences, University of Eastern Finland, Kuopio, Finland
e-mail: oksana.skaldina@uef.fi

© Springer Nature Switzerland AG 2019
K. K. Kesari (ed.), *Networking of Mutagens in Environmental Toxicology*, Environmental Science,
https://doi.org/10.1007/978-3-319-96511-6_7

137

1 Introduction

Climate change, forest clearing, intensive agriculture, global trade and pollution are main man-induced defendants of life transformations on the Earth. Afore global environmental change understanding of living beings' adaptability to various environmental stresses is crucial for decision making policy, environmental management and environmental law (Chasek 2018). Despite vast scientific studies and public efforts, environmental pollution remains among key environmental concerns globally. Such pollutants as chlorofluorocarbons (CFCs), organochlorine (OC), polychlorinated biphenyls (PCBs), perfluorinated compounds (PFCs), organotins (OTs) and heavy metals (HM), released from industry and transport, penetrate into aquatic and terrestrial ecosystems and cause acute or chronic threats to biota (Mateo et al. 2016). However, the lack of knowledge about direct effects of pollution on different biological compartments of terrestrial ecosystems requires further analyses on the topic.

Heavy metal pollution may affect all components of terrestrial ecosystems, from soil and microbes to vertebrates (Gall et al. 2015). As trace elements, the heavy metals such as cobalt (*Co*), copper (*Cu*), iron (*Fe*), manganese (*Mn*), molybdenum (*Mo*), selenium (*Se*) and zinc (*Zn*) are essential micronutrients for plants, animals and humans. While some of the other elements, like arsenic (*As*), cadmium (*Cd*), chromium (*Cr*), mercury (*Hg*), lead (*Pb*) and nickel (*Ni*) have no useful biological function and might cause toxic effects even at low concentrations (He et al. 2005).

Revealing biological responses towards heavy metal pollution in living organisms, particularly in those, which are of great ecological and economic importance, and developing practical biomonitoring tools, may effectively facilitate progress in modern terrestrial ecotoxicology (Skaldina and Sorvari 2017). Especially since *"the concept of biological monitoring, based on the study of the biological response of organisms to pollutants, termed biomarkers, is today well established"* (Romeo and Giamberini 2013). This is practically important for further development of ecotoxicology, aiming to discover causal linkages between the source of pollution and range of its biological effects with the purpose to reduce an impact via different kinds of legislative, social and environmental interventions (Elliott et al. 2011).

2 Economic Importance of Insects

Insects represent one of the most diverse groups of living organisms, with estimated possible number of species up to 10 million (Gaston 1991). They are important drivers of key ecosystem processes, which are ultimately mediated through interactions between all ecosystem components. Even though cumulative biomass of insects contributes less to global carbon and nutrient cycling than, for example, the total biomass of plants and microbes (Yang and Gratton 2014).

Insects provide substantial provisioning, regulating, supporting and cultural ecosystem services such as plant pollination and food provisioning, medicine services, biological control, recycling organic matter, soil nutrient and fertility regulation, biodiversity protection, bioindicators and conservation tools, religion and spiritual value and cultural heritage (Noriega et al. 2018). So what kind of beneficial tasks do terrestrial insects perform? They pollinate plants and disperse seeds, protect crop and control pests, maintain soil structure and cycle nutrients, maintain food webs and favor ecosystem health (Scudder 2017). Economic value of the ecosystem services such as dung burial, pest control, pollination and wildlife nutrition, provided by wild not domesticated or mass-reared insects in the USA is estimated as $57 billions per year (Losey and Vaughan 2006).

Economically important, beneficial insects belong to various systematic categories. For instance, bees, hover flies, lacewings, predatory bugs, caterpillar parasitoids and ants. However, insects are responsible for numerous adverse environmental and ecological concerns as well. Many insect species, belonging to the orders Lepidoptera, Coleoptera, Orthoptera and Hymenoptera (Van Emden and Wearing 1965; Del Toro et al. 2012), are serious agricultural pests. Therefore, economic role of various insects should be thoroughly considered in different types of environmental surveys.

3 Current State of Environmental Heavy Metal Pollution and General Ecotoxicological Effects in Terrestrial Insects

Although, current levels of industrial heavy metal pollution have been decreasing in European Union (Tóth et al. 2016) the opposite tendency is occurring in China (Li et al. 2014) and several other eastern countries (Järup 2003).

Different metal ions are vital for normal physiological processes in small concentrations, however, poisonous in higher amounts. Therefore, organisms should possess proper regulatory mechanisms for metal uptake, assimilability and excretion. Insects' adaptability to heavy metal pollution depends on their dispersal capacities, as mobile organisms may drift to better habitats, while sessile species should derive phenotypic and genetic adaptations (Merritt and Bewick 2017). Profound investigation of heavy-metal tolerance (*Cd*) in soil-living springtail *Orchesella cincta* revealed that in metal-exposed field populations the metallothionein gene is overexpressed in this species (Janssens et al. 2009). Therefore, it was suggested that *cis*-regulatory change of genes, engaging into cellular stress response, might be significant for the evolution of tolerance mechanisms towards heavy metal pollution in insects.

Heavy metals induce diverse ecotoxicological effects on terrestrial insects and affect some taxa more than the others. Thus, knowledge on the responses in few species cannot be generalized over different insect orders.

4 Ecotoxicological Effects of Heavy Metals on Pollinators

Pollinators is, probably, one of the most important functional group of insects, providing crucial ecosystem services for crops and wild plants. Pollinator decline and consequent loss of pollination services will rapidly result in adverse effects on crop production, food security and human welfare (Potts et al. 2010). Both honey bees (*Apis mellifera*) and bumble bees (*Bombus* sp.), which are key pollinators in diverse terrestrial ecosystems, accumulate heavy metals in their bodies (Lindqvist 1993; Fakhimzadeh and Lodenius 2000; Szentgyörgyi et al. 2011; Satta et al. 2012). Meindl and Ashman (2013) suggested that soil metals, especially *Ni*, can impair foraging behaviour of bumble bees in polluted areas. In their experimental study, it was shown that nectar solutions with elevated levels of nickel were visited for shorter time periods than the control ones.

Besides honey bees and bumble bees, heavy metals might affect other pollinators. However, ecotoxicological effects of heavy metals on wild bees are poorly understood. Szentgyörgyi et al. (2017) failed to reveal any association between heavy metal contamination and developmental instability, measured as fluctuating asymmetry (FA) of forewing. Therefore, more studies are needed to understand the effect of heavy metal pollution on various pollinating insects.

5 Ecotoxicological Effects of Heavy Metal Pollution in Important Predatory Insects

Many insect species are important natural predators in food chains and crucial biological control agents in pest management programs. Ants, wasps, bugs and lacewings comprise the most economically valuable groups of predatory insects.

Being important predatory insects, ants (Formicidae) can tolerate heavy metal pollution quite well (Folgarait 1998; Grześ 2010). Nevertheless, it can induce ecotoxicological responses related to morphology or physiology and alter behaviour in this group of insects. Recently we have found that hairy wood ant *Formica lugubris* respond to heavy metal contamination with decreased body mass and head melanin-based colouration (Skaldina et al. 2018). Grześ et al. (2015b) found that the size-distribution of workers in black garden ant *Lasius niger* colonies in a polluted area was biased to small workers.

In heavily polluted areas, ants demonstrate disturbed immune responses and therefore may be subject to higher risk of infections (Sorvari et al. 2007). Physiological ecotoxicological responses in ants to heavy-metal pollution correlates with exposure time. Under experimental set-up it was shown that prolonged feeding with *Cd* and *Hg* contaminated food diminished their ability to maintain proper energetic balance and resulted in decreased activity of several vital enzymes (Migula et al. 1997). It was also discovered that heavy metal pollution can decrease normal aggressive behaviour between workers of different ant colonies in *Formica aquilonia* (Sorvari and Eeva 2010). Aggressive territorial behaviour is needed to maintain proper population structure and colony integrity. Potentially, this behavioural alteration might modify structure of invertebrate communities in boreal and temperate forests, as *F. aquilonia* is an ecologically dominant species in these areas (Sorvari and Eeva 2010). No signs of developmental instabilities have been revealed in ants. Red wood ants *Formica pratensis* and yellow meadow ants *Lasius flavus*, inhabiting heavily metal-contaminated sites in Austria and Poland did not show signs of fluctuating asymmetry (Rabitsch 1997; Grześ et al. 2015a). It was suggested that some ant species might possess effective mechanisms of heavy metal regulation (Grześ 2009), therefore not every ant species are suitable for monitoring heavy metal pollution.

Ecotoxicological responses of wasps to heavy metal pollution are practically unrevealed. To our knowledge, one study about wasps as bioindicators, addressed levels of pollutants in wasps' feces (Urbini et al. 2006) and the other one revealed elevated lead concentrations in midgut epithelium of paper wasps (Polidori et al. 2018). Our results revealed that common wasp *Vespula vulgaris* accumulates heavy metals such as *As*, *Co*, *Cu*, *Fe*, *Ni*, *Pb*, *Sr* and *Zn*, and from those the levels of *As*, *Cd*, *Cu* and *Pb* decreased with an increase in distance from Harjavalta *Cu-Ni* smelter (Skaldina et al. unpublished). Biochemical and physiological mechanisms of such responses to heavy metal pollution in wasps should be studied further.

Beetles (Coleoptera) are generally less capable for heavy metal accumulation in compare with some other invertebrates (Heikens et al. 2001; Butowski 2011). The majority of studies about ecotoxicological effects of metals in beetles were done with the model species and very limited information is available about the other .

6 Heavy Metals and Pest Insects

Aphids and butterfly larvae are among the most damaging pest insects. Heavy metal pollution may lead to morphological and physiological alterations in these organisms as well. Görür (2006) revealed that metals, accumulated in host plant, might result in expression of morphological traits (size of various body parts) in aphids. It was shown that cadmium significantly affects life-history traits and metabolic enzymes in cotton bollworm *Helicoverpa armigera*. Experimental setup with an increased *Cd*-supplemented diet resulted in belated larval development, decreased survival rate, decreased female fecundity and altered enzymes (GST, CarE, P450 and AChE) activity in this species (Zhan et al. 2017). Sun et al. (2011) found immune sensitivity of important Asian Lepidopteran pest *Spodoptera litura* to nickel pollution, and it was dependent on the *Ni* concentrations and periods of exposure.

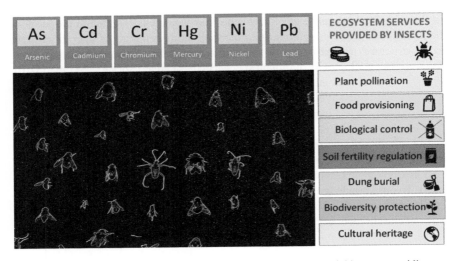

Fig. 1 Effects of toxic heavy metals on economically important terrestrial insects, providing crucial ecosystem services, may induce morphological, physiological and behavioural alterations and threaten biodiversity with further consequences to food security and human welfare

7 Conclusions

To summarize, heavy metal pollution may induce different morphological, physiological and behavioural alterations in economically important insects, such as pollinators, predators and pests. Current knowledge about ecotoxicological effects of heavy metals in important terrestrial insects is quite scarce. Most of studies on the topic addressed model species with no or little significant consideration of the ecological or economical role of the organisms. Ecotoxicological effects of heavy metal pollution in economically important insects should receive more scientific attention, as it has direct consequences to food security, human welfare and health (Fig. 1).

References

Butowski R (2011) Heavy metals in carabids (Coleoptera, Carabidae). ZooKeys 100:215–222

Chasek PS (2018) Global environmental politics. Routledge, New York, p 482

Del Toro I, Ribbons RR, Pelini SL (2012) The little things run the world revisited: a review of ant-mediated ecosystem services and disservices. Myrmecol News 17:133–146

Elliott JE, Bishop CA, Morrissey CA (eds) (2011) Wildlife ecotoxicology: forensic approaches. Springer, New York, 468 pp

Fakhimzadeh K, Lodenius M (2000) Heavy metals in Finnish honey, pollen and honey bees. Apiacta 35(2):85–95

Folgarait PJ (1998) Ant biodiversity and its relationship to ecosystem functioning: a review. Biodiv Conserv 7:1221–1244

Gall JE, Boyd RS, Rajakaruna N (2015) Transfer of heavy metals through terrestrial food webs: a review. Environ Monit Assess 187(4):201

Gaston KJ (1991) The magnitude of global insect species richness. Conserv Biol 5:283–296

Görür G (2006) Effects of heavy metal accumulation in host plants to cabbage aphid (*Brevicoryne brassicae*)—morphology. Ekológia (Bratislava) 25(3):314–321

Grześ IM (2009) Cadmium regulation by *Lasius niger*: a contribution to understanding high metal levels in ants. Insect Sci 16:89–92

Grześ IM (2010) Ants and heavy metal pollution—a review. Eur J Soil Biol 46:350–355

Grześ IM, Okrutniak M, Szpila P (2015a) Fluctuating asymmetry of the yellow meadow ant along a metal-pollution gradient. Pedobiologia 58:195–200

Grześ IM, Okrutniak M, Woch MW (2015b) Monomorphic ants undergo within-colony morphological changes along the metal-pollution gradient. Environ Sci Pollut Res 22:6126–6134

He ZL, Yang XE, Stoffella PJ (2005) Trace elements in agroecosystems and impacts on the environment. J Trace Elem Med Biol 19(2–3):125–140

Heikens A, Peijnenburg WJGM, Hendriks AJ (2001) Bioaccumulation of heavy metals in terrestrial invertebrates. Environ Pollut 113:385–393

Janssens TKS, Roelofs D, Van Straalen NM (2009) Molecular mechanisms of heavy metal tolerance and evolution in invertebrates. Insect Sci 16:3–18

Järup L (2003) Hazards of heavy metal contamination. Brit Med Bullet 68:167–182

Li Z, Ma Z, van der Kuijp TJ, Yuan Z, Huang L (2014) A review of soil heavy metal pollution from mines in China: pollution and health risk assessment. Sci Total Environ 468–469:843–853

Lindqvist L (1993) Individual and interspecific variation in metal concentrations in five species of bumblebees (Hymenoptera: Apidae). Environ Entomol 22(6):1355–1357

Losey JE, Vaughan M (2006) The economic value of ecological services provided by insects. Bioscience 56(4):311–323

Mateo R, Lacorte S, Taggart MA (2016) An overview of recent trends in wildlife ecotoxicology. Current trends in wildlife research. In: Mateo R, Arroyo B, Garcia J (eds) Current trends in wildlife research. Wildlife Research Monographs, vol 1. Springer, Cham, pp 125–150

Meindl GA, Ashman T-L (2013) The effects of aluminium and nickel in nectar on the foraging behaviour of bumblebees. Environ Pollut 177:78–81

Merritt TJS, Bewick AJ (2017) Genetic diversity in insect metal tolerance. Front Genet 8:172

Migula P, Głowacka E, Nuorteva SL, Nuorteva P, Tulisalo E (1997) Time related effects of intoxication with cadmium and mercury in the red wood ants (*Formica aquilonia*) Ecotoxicology (6):307–320

Noriega JA, Hortal J, Azcáratec FM, Berg MP, Bonada N, Briones MJI, Del Toro I, Goulson D, Ibanez S, Landis DA, Moretti M, Potts SG, Slade EM, Stout JC, Ulyshen MD, Wackers FL, Woodcock BA, Santos AMC (2018) Research trends in ecosystem services provided by insects. Basic Appl Ecol 26:8–23

Polidori C, Pastor S, Jorge S, Pertusa J (2018) Ultrastructural alterations of midgut epithelium, but not greater wing fluctuating asymmetry, in paper wasps (*Polistes dominula*) from urban environments. Microsc Microanal 24(02):183–192

Potts SG, Biesmeijer JC, Kremen C, Neumann P, Schweiger O, Kunin WE (2010) Global pollinator declines: trends, impacts and drivers. Trends Ecol Evol 25(6):345–353

Rabitsch WB (1997) Levels of asymmetry in *Formica pratensis* Retz. (Hymenoptera, Insecta) from a chronic metal-contaminated site. Environ Toxicol Chem 16(7):1433–1440

Romeo M, Giamberini L (2013) History of biomarkers. In: Amiard-Triquet C, Amiard J-C, Rainbow PS (eds) Ecological biomarkers: indicators of ecotoxicological effects. CRC Press, Boca Raton, pp 15–43

Satta A, Verdinelli M, Ruiu L, Buffa F, Salis S, Sassu A, Floris I (2012) Combination of beehive matrices analyses and ant biodiversity to study heavy metal pollution impact in a post-mining area (Sardinia, Italy) Environ Sci Pollut Res 19:3977–3988

Scudder GGE (2017) The importance of insects. In: Foottit RG, Adler PH (eds) Insect biodiversity: science and society, volume I, 2nd edn. Wiley-Blackwell, 904 pp

Skaldina O, Sorvari J (2017) Biomarkers of ecotoxicological effects in social insects. In: Kesari K (ed) Perspectives in environmental toxicology. Springer International Publishing, Switzerland, pp 203–214

Skaldina O, Peräniemi S, Sorvari J (2018) Ants and their nests as indicators of industrial heavy metal pollution. Environ Pollut 240:574–581

Sorvari J, Eeva T (2010) Pollution diminishes intra-specific aggressiveness between wood ant colonies. Sci Total Environ 408(16):1389–1392

Sorvari J, Rantala LM, Rantala MJ, Hakkarainen H, Eeva T (2007) Heavy metal pollution disturbs immune response in wild ant populations. Environ Pollut 145(1):324–328

Sun HX, Danq Z, Xia Q, Tanq WC, Zhanq GR (2011) The effect of dietary nickel on the immune responses of *Spodoptera litura* Fabricius larvae. J Insect Physiol 57(7):954–961

Szentgyörgyi H, Blinov A, Eremeeva N, Luzyanin S, Grześ IM, Woichiechowski M (2011) Bumblebees (Bombidae) along pollution gradient-heavy metal accumulation, species diversity, and *Nosema bombi* infection level. Polish J Ecol 59(3):599–610

Szentgyörgyi H, Moron D, Nawrocka A, Tofilski A, Woyciechowski M (2017) Forewing structure of the solitary bee *Osmia bicornis* developing on heavy metal pollution gradient. Ecotoxicology 26:1031–1040

Tóth G, Hermann T, Da Silva MR, Montanarella L (2016) Heavy metals in agricultural soils of the European Union with implications for food safety. Environ Internat 88:299–309

Urbini A, Sparvoli E, Turilazzi S (2006) Social wasps as bioindicators: a preliminary research with *Polistes dominulus* (Hymenoptera, Vespidae) as a trace metal accumulator. Chemosphere 64(5):697–703

Van Emden HF, Wearing CH (1965) The role of the aphid host plant in delaying economic damage levels in crop. Annal Appl Biol 56(2):323–324

Yang LH, Gratton C (2014) Insects as drivers of ecosystem processes. Curr Opin Insect Sci 2:26–32

Zhan H, Zhang J, Chen Z, Huang Y, Ruuhola T, Yang S (2017) Effects of Cd^{2+} exposure on key life history traits and activities of four metabolic enzymes in *Helicoverpa armigera* (Lepidopteran: Noctuidae). Chem Ecol 33(4):325–338

Contamination Links Between Terrestrial and Aquatic Ecosystems: The Neonicotinoid Case

Victor Carrasco-Navarro and Oksana Skaldina

Abstract Current rates of economic development are interrelated with an increase in environmental pollution. Among different contamination agents, modern insecticides such as neonicotinoids (NNIs) require precise attention in evaluation of losses and benefits. NNIs is relatively new class of systemic insecticides, being in use for about 20 years and embracing around 25% of global pesticide market. Currently there are several methods to apply NNIs to plants such as foliar sprays, soil drenches and seed treatments, and in recent years there has been a global shift towards seed treatment (seed dressing) rather than aerial spraying. The discovery of NNIs was considered as a milestone in the research on insecticides. Possessing chemical structure similar to nicotine and acting as agonists at insects' acetylcholine receptors, NNIs demonstrate selective toxicity to invertebrates versus vertebrates. In addition, toxicity of NNIs in mammals is between one to three orders of magnitude lower than the toxicity caused by their predecessors: organophosphates, carbamates and pyrethroids. However, NNIs are mobile contaminants that can be transferred from plants to soils and water and induce diverse array of toxic effects in non-target organisms, even affecting animals not in contact with them directly. Surface- and groundwater may also act as vector for the transport of NNIs to untreated locations. The presence of NNIs in water bodies might facilitate their uptake by non-target plants present in littoral and riparian zones, with the potential threat to herbivorous insects. Leaching of NNIs to groundwater may imply their further distribution to other matrices, potentially leading to undesirable environmental issues. Pollinators and aquatic insects appear to be especially susceptible to these insecticides and chronic sublethal effects tend to be more prevalent than acute toxicity. Although a complete knowledge of the fate of NNIs in the environments is missing, authorities are starting to react to the threat they pose by limiting their use and application. Relevant improvements have been made in the field of the toxicity to non-target organisms. Studies that include factors such as mixture toxicity, field or semi-field exposures can make significant contribution to the further evaluating of costs-benefits of neonicotinoids.

V. Carrasco-Navarro (✉) · O. Skaldina
Department of Environmental and Biological Sciences, University of Eastern Finland, Kuopio 70211, Finland
e-mail: victor.carrasco.navarro@uef.fi

© Springer Nature Switzerland AG 2019
K. K. Kesari (ed.), *Networking of Mutagens in Environmental Toxicology*, Environmental Science,
https://doi.org/10.1007/978-3-319-96511-6_8

Keywords Agricultural pollution · Aquatic insects · Pollinators · Predators · Neonicotinoids · Non-target organisms · Systemic insecticides

1 Introduction

Current rates of industrial and agricultural development have inevitable side effects and one of those is environmental contamination (Grossman and Kruger 1995). Persistent organic pollutants (POPs), petroleum hydrocarbons, heavy metals, plastic pollutants and pesticides are some of the most common contaminants affecting terrestrial and aquatic ecosystems in a daily routine. Their massive or continuous release can provoke ecological disasters, causing acute toxicity to many different organisms. However, inconspicuous, hidden or extended consequences can be even more dangerous. Not resulting in immediate alarming effects, they may go unnoticed for a long time before the critical moment comes. In addition, complex interactions between climate change and pollution can alter physical and chemical stressors, starting to be even more problematic for the organisms, living at the edge of their tolerance (Noyes et al. 2009). Finally, the ongoing environmental contamination and related ecosystem change lead to widespread species extinction and biodiversity loss (Butchart et al. 2010; Hooper et al. 2012).

Terrestrial and aquatic ecosystems are closely interrelated. One of the major land-water linkages in the biosphere is gravitational movement of material in drainage waters (Likens and Bormann 1974). Some contaminants entering soil may cause pronounced effects to organisms living in water. Holistic approach regarding land-air-water interactions is required for intelligent management of landscapes.

Regarding pesticide pollution, the case of neonicotinoid insecticides is one of the most pressing issues nowadays. Neonicotinoids (new nicotine-like insecticides—NNIs) are a family of toxic substances, including imidacloprid, acetamiprid, dinotefuran, nitenpyram, thiamethoxam, thiacloprid and clothianidin, are used to kill agricultural pest insects (Simon-Delso et al. 2015). Also, they are in use in veterinary medicine for controlling parasitic insects such as ticks or fleas. Neonicotinoids are relatively new class of systemic insecticides, being in active use for only two decades. NNIs possess chemical structure similar to nicotine and act as agonists at insects' acetylcholine receptors (nAChRs), exhibiting selective toxicity to invertebrate versus vertebrate species (Matsuda et al. 2001). The approval of the first neonicotinoid, imidacloprid, was granted in 1994 in USA and in 2005 in the EU. During the following years, the other neonicotinoids were developed and approved to be used in the market. These were nitenpyram, thiamethoxam, clothianidin, dinotefuran, thiacloprid and acetamiprid.

Many recent studies confirmed toxic effects of NNIs to non-target organisms (Beketov and Liess 2008; Lever et al. 2014; Addy-Orduna et al. 2019). Meanwhile, it is still necessary to summarise and understand general linkages of neonicotinoid pollution between terrestrial and aquatic ecosystems. Therefore, here we aimed to review current state of knowledge about neonicotinoid contamination in terrestrial and aquatic ecosystems and the toxic effects of NNIs to non-target organisms, living in these ecosystems.

2 History of Neonicotinoids

The discovery of NNIs was considered as a milestone in the research on insecticides (Tomizawa and Casida 2011). Their solubility in water, together with their low toxicity to humans and other species of mammals made them a great choice among the available plant protection products. The insecticide market slowly switched from organophosphates, carbamates and pyrethroids to NNIs (Wood and Goulson 2017). Comparing the oral LD_{50} in rats (available in Yu 2015), it can be concluded that the toxicity of NNIs is between one and three orders of magnitude lower than the toxicity caused by their predecessors, such as organophosphates, carbamates and pyrethroids (Yu 2015).

However, NNIs were not as safe as it seemed for other organisms. Already during the registration process, imidacloprid was found highly toxic to three species of bees, with a LD_{50} of 0.0439 µg/bee (USEPA 1992). In 2008, an accidental release of clothianidin affected 11,000 hives and resulted in a massive death of bees in Germany (BVL 2008). The EU commission requested a conclusive assessment to the European Food Safety Authority (EFSA) on the risk of three NNIs (imidacloprid, thiamethoxam and clothianidin) to bees (EFSA 2012).

EFSA presented evidence of sublethal toxic effects in bees resulting from the exposure to these insecticides (EFSA 2013a, b, c). As the EU member states did not come to an agreement about the banning of these substances (some state members did not agree and some others such as Finland abstained), the European commission adopted a proposal (Regulation No. 485/2013) that prompted the restriction in the use of three NNIs for the period of two years, until new scientific evidence was gathered. The restriction started on December 1st, 2013, and it applied to three members of the neonicotinoid family, imidacloprid, clothianidin and thiamethoxam. Additionally, during the following years, several emergency authorizations were granted to a total of seven member states despite to the ban (e.g.: EFSA 2018a, b). Authorizations were granted due to the lack of alternative measures or products against specific pests.

In addition, the five NNIs approved for use in the EU were selected in 2016 to complete the First Watch list for emerging water pollutants (European Commission 2015). EFSA examined the new evidence related to the risk assessment of the three NNIs involved and after long deliberations, the EU commission implemented the regulations that ban the outdoor use of imidacloprid, clothianidin and thiamethoxam in May 2018 (Regulations (EU) 2018/783, 2018/784 and 2018/785, respectively). It must be highlighted that the regulations do allow the use of the three compounds indoors (e.g.: in greenhouses), what may protect bee populations but also may cause NNIs to leach to water bodies. It is also important to remark that the ban does not affect the use of other NNIs such as acetamiprid and thiacloprid. At first these two compounds were found to be less toxic to bees than imidacloprid, thiamethoxam and clothianidin (EFSA Panel PPR 2012), but they are highly toxic to freshwater invertebrates (Raby et al. 2018a) and therefore a threat for the future of aquatic organisms and their surrounding ecosystems.

In the US, the environmental protection agency (US-EPA), released NNIs assessments for public comment in 2017 and will come to a definitive conclusion during 2019, aiming at reducing risk. The NNIs involved in these procedures are imidacloprid, thiamethoxam, clothianidin, dinotefuran and acetamiprid. It is worth noting that thiacloprid permission was voluntarily cancelled by the manufacturer already in 2014.

3 Neonicotinoids in the Environments

Neonicotinoids are systemic insecticides: the active compounds distribute to all plant tissues and therefore provide a more complete protection to pests that would feed on any part of the plant. Due to the unique properties of NNIs such as increased intrinsic acute and residual activity for many agricultural pests (whiteflies, Colorado potato beetles, aphids etc.) and their systemic properties, they are applicable to many different crops: rape, corn, cotton, potatoes, sugar beet, tobacco, cereals, pome fruits (Jeschke and Nauen 2008).

Currently there are several methods to apply NNIs to plants, including foliar sprays, soil drenches and seed treatments (Bonmatin et al. 2015). In recent years there was a global shift towards widespread application of these insecticides as a seed treatment rather than aerial spraying (Hladik et al. 2018). The so called "seed dressing" initiates the protection of the plant at the seed stage and, due to the solubility of NNIs, helps to the distribution of the insecticides to all plant tissues. Flowers and pollen would therefore contain the applied neonicotinoid(s) even if application methods such as foliar spray are avoided. Seed dressing additionally minimizes the loading of pesticides to the surrounding environment and reduces the occupational exposure of farmers compared to spraying or foliar treatments.

When the plant has flowered, NNIs are distributed throughout all the plant tissues. Different species of invertebrate and vertebrate pollinators are in contact with NNIs through flowers and pollen and may uptake them. Neonicotinoids are present at every parts of plants growing from treated seeds: in stem, leaves, nectar and pollen, and it is generally assumed that from 2 to 20% of pesticide's coating is absorbed by plants' tissues (Alford and Krupke 2017; Hladik et al. 2018).

There are several routes of environmental exposure of NNIs from treated seeds to terrestrial and aquatic ecosystems (Fig. 1). Seeds can be ingested by birds and some small terrestrial mammals. NNIs are highly persistent in soils and are characterized by a high runoff and leaching potential to surface and groundwater (Bonmatin et al. 2015). Penetrating aquatic environments, they become a threat to larvae of aquatic insects and fish.

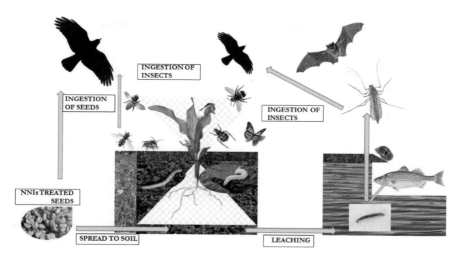

Fig. 1 Links of neonicotinoid contamination between terrestrial and aquatic ecosystems

4   NNIs Contamination: Links Between Ecosystems

Despite the good intentions of the seed dressing treatments, up to 95% of the NNIs that are coating the seeds leach to the surrounding soils (Sur and Stork 2003). In soils, NNIs are persistent, with DT_{50}'s reaching thousands of days (Goulson 2013). If the DT_{50} reach over one year, the compounds are accumulating in the soil and most likely their concentrations increase with time. Neonicotinoids are prone to leaching if the right conditions are met. Usually, rainfalls (Hladik et al. 2014) or even melting snow (Main et al. 2016) can provoke a rise in neonicotinoid concentrations in the surrounding water bodies. Thus, it has been common to find NNIs in water bodies all around the world in concentrations ranging from the low ng L^{-1} to the hundreds μg L^{-1} (Morrissey et al. 2015), and often surpassing the concentrations set as quality guidelines for e.g.: imidacloprid. In addition to the findings in surface water bodies, NNIs have been detected also in groundwater (Bradford et al. 2018).

Surface- and groundwater may act as vector for the transport of NNIs to untreated locations. Two key aspects related to this transport have not been extensively investigated and are important when considering the whole neonicotinoid cycle since they enter the environment. First, the presence of NNIs in water bodies might facilitate their uptake by non-target plants present in littoral and riparian zones, with the potential threat to herbivorous insects. A similar scenario was first presented by Goulson (2013), who suggested that NNIs might be available from soils to non-target flora in areas near treated fields, what has been recently reviewed in Wood and Goulson (2017). Second, the leaching of NNIs to groundwater may imply their further distribution to other matrices, potentially leading to undesirable environmental issues (Huseth and Groves 2014). If groundwater is used as irrigation, NNIs may recirculate to the same crop they were originally applied to (Huseth and Groves 2014).

It is still unknown whether lotic waters (surface- and groundwaters) can transport NNIs to areas far away from the agricultural fields where they were originally applied to. In addition, it would be interesting to know whether non-target plants or organisms uptake NNIs from these lotic water bodies.

Most likely, pollinators are the vector of transport of NNIs to their predators, as for example thiacloprid and imidacloprid have been found in the blood of the European honey buzzard (Byholm et al. 2018), as one of its major food sources are pollinating bumble bees. The transfer to honey buzzards may be reinforced by the fact that bees that have uptaken NNIs are less likely to avoid predators (Tan et al. 2014).

Contrarily to what occurs with other hydrophobic contaminants such as PCBs, it is not expected that the concentrations of neonicotinoids increase along the trophic chain (biomagnification), due to their solubility in water and therefore their excretion in urine. However, the finding of imidacloprid and thiacloprid in this long distance migrating raptor opens more questions about the presence of NNIs in other animals and the toxic effects that they may cause. Unfortunately, the aforementioned study is not the only one that has recently reported the presence of NNIs in birds. Additionally to the honey buzzard, NNIs have been found in other birds such as the Eurasian eagle owl (Taliansky-Chamudis et al. 2017), hummingbirds (Bishop et al. 2018) and quail (Turaga et al. 2016). The ingestion of NNIs by birds may result in toxic effects. It has been found that the South America eared dove would need to eat as less as 1.7 g of seeds to reach the LD_{50} of imidacloprid (Addy-Orduna et al. 2019). Importantly, a recent article reported a reduction in the migration ability of songbirds (Eng et al. 2017), finding that opens questions about the effects of NNIs in other migrating species such as the European honey buzzard.

Additional to the toxic effects caused by direct ingestion or contact with NNIs present in seeds, prey or pollen, indirect effects caused by other factors may also affect bird populations (Gibbons et al. 2015). Factors as declining populations of invertebrate prey are important when considering the whole consequences in the use of NNIs and other pesticides. A decrease in insectivorous birds has been associated with higher concentrations of imidacloprid in surface water of The Netherlands (Hallmann et al. 2014), what constitutes a dramatic example of indirect effects to entire bird populations.

5 Ecotoxicological Effects of NNIs in Non-target Species

Neonicotinoids specifically bind and continuously activate the insect nicotine acetyl-choline receptor (nAChR), causing a series of disorders and finally leading to the death of an insect (Yu 2015). Certainly, there is no specific distinction between the binding capacities of NNIs to the nAChR in target vs. non-target insects. Thus, the high sensitivity of most insects to NNIs does not come as a surprise. The main use of NNIs has led to widespread detection of NNIs in the environment (in soil, water, pollen or honey). Pollinators and aquatic insects appear to be especially susceptible to these insecticides and chronic sublethal effects tend to be more prevalent than acute toxicity.

5.1 Neonicotinoids and Non-target Terrestrial Insects: Threats to Pollinators

When comparing agricultural pollution to other pollution types, environmental contamination by pesticides might appear relatively insignificant, however pesticide residues in soils might be toxic to soil microbial and invertebrate faunas (Iyaniwura 1991). Those insects, living in soil permanently or partly during some their stages facing this type of exposure. There are many of such examples, including ground-nesting bumble bees and wild solitary bees, spider wasps, larvae of predatory beetles or hoverflies. Many of those are important pollinators and natural predators, helping to control and reduce populations of pest insects. One extreme example of the threats posed by pesticides was the case of bees' mass-dying occurred close to corn fields during sowing NNIs-treated seeds. That was due to acute intoxication via exposure to the dust clouds near sowing machines (Girolami et al. 2012). However, the effects of such sowing techniques on many other wild beneficial insects remains unknown.

Ongoing chemicalization of terrestrial ecosystems is one of the most severe danger for overall biodiversity and for resilience in important ecosystem services (ES), provided by insects. Insects are major agents for plant pollination, which is one of the most essential ES (Noriega et al. 2018). Indeed, the value pollination is widely accepted in financial, food security and health terms, as insect pollination services represent 9.5% of global crop production value (Gallai et al. 2009). Substantial loss of pollinators has been documented in many regions of the globe (Potts et al. 2010; Cameron et al. 2011; Lever et al. 2014). A horizon scan approach determined novel NNIs to be among six major issues of high priority, which will remain significant threats for pollinators in the nearest future (Brown et al. 2016).

Both commercial and wild pollinators such as honeybee, wild bees, bumblebees, beetles, wasps, ants and butterflies are in the high risk-zone. Honey bee (*Apis mellifera*) is the most studied non-target terrestrial invertebrate (Pisa et al. 2015). Because of the specific metabolic routes and target links to nervous system NNIs have direct effects on learning capacities, memory and behavior. Wild pollinators are also especially susceptible to neonicotinoid pesticides, which induce chronic sub lethal effects rather than acute toxicity (Hladik et al. 2018). Neonicotinoid pesticides are spreading from agricultural areas to neighboring wildflowers and make greater impact on wild pollinators than it was initially assumed (Botías et al. 2016). Because NNIs are systemic pesticides, pollinating insects are exposed to small amounts of insecticides each time, when they feed on pollen or nectar of treated plants (Godfray et al. 2014).

5.2 NNIs and Non-target Aquatic Organisms

Once the presence of NNIs in diverse aquatic ecosystems was confirmed, the question was whether the levels of the NNIs reported were a threat to the species living in these ecosystems. In order to comply with the protection of the aquatic environment and its organisms against chemical threats, threshold concentrations of NNIs were set by environmental agencies in different parts of the World. For imidacloprid, the threshold concentrations are as low as 0.23 μg L^{-1} for Canadian freshwaters (CCME 2007) and an annual average of 0.0083 μg L^{-1} for Dutch freshwaters (maximum acceptable concentrations of 0.2 μg L^{-1}; RIVM 2014). In countless occasions these values are exceeded, with the consequent risk to populations of aquatic organisms.

The values for freshwater in The Netherlands are more complete, since they distinguish average and maximum concentrations. In the real environment, variable concentrations are usually found, especially in lotic water bodies near agricultural areas. Neonicotinoids may be detected in pulses after rain events in these water bodies (Hladik et al. 2014; Beketov and Liess 2008), due to their leaching potential. In lentic water bodies such as wetlands, low but more constant concentrations of NNIs may be the dominant trend (Maloney et al. 2018). Therefore, the patterns of exposure of organisms dwelling in lentic and lotic may be completely different (Raby et al. 2018a).

It has been proved that aquatic insects are more sensitive to NNIs than other taxa (Beketov et al. 2008; Raby et al. 2018a, b) in both laboratory and mesocosm experiments. Morrissey et al. (2015) comprehensively reviewed the toxicity of aquatic invertebrates to NNIs and determined that the orders Ephemeroptera (mayflies), Trichoptera (caddisflies) and Diptera (flies) are, in that order the most sensitive aquatic insects. The responses of different species over short or long-term studies may vary dramatically and a single pulse of one neonicotinoid can alter the abundance and taxa richness of invertebrates in streams (Beketov et al. 2008). Interestingly, recovery of some species of invertebrates occurs weeks after the contamination episode or under low concentrations, even after a pronounced initial decline (Beketov et al. 2008; Rico et al. 2018; Pickford et al. 2018). Multivoltine insects such as mayflies and chironomids can recover from stress episodes. The fact that these species have a short life cycle and produce several generations per year is an advantage compared to the univoltine species, which larvae develops at a slower pace (Beketov et al. 2008; Rico et al. 2018). Additionally, NNIs have been found to affect reproduction (Raby et al. 2018b) and metamorphosis (Raby et al. 2018b), for example ecdysis (Cavallaro et al. 2018) in emerging insects, which are key species in connecting aquatic and terrestrial ecosystems (Baxter et al. 2005), as we have suggested above.

At the molecular level, imidacloprid caused oxidative stress in the amphipod *Gammarus fossarum* (Malev et al. 2012). The concentrations at which the toxic effects were observed were over 100 μg L^{-1}, what proves the decreased sensitivity in amphipods compared to insects.

Regarding to aquatic vertebrates, some signs of immuno- and genotoxicity have been reported in fish (Hong et al. 2018; Iturburu et al. 2017; Velisek and Stara 2018).

Also, the responses of Wood frogs to predation were altered in frogs exposed to imidacloprid as tadpoles (Lee-Jenkins and Robinson 2018). However, the concentrations at which toxic effects are found are several times higher than the concentrations at which NNIs are toxic to insects. In addition, positive results are usually found at concentrations not environmentally relevant. However, Vieira et al. (2018) found a significant increase in DNA damage in erythrocytes of *Prochilodus lineatus* at 1.25 μg L^{-1} of imidacloprid and Velisek and Stara (2018) found changes in the levels of antioxidant enzymes in the early life stages of the common carp at 4.5 μg L^{-1}.

It is plausible that NNIs are genotoxic or immunotoxic at high concentrations. However, this would not be evident in more sensitive organisms, as at high concentrations mortality would occur before any toxicity is observed at the molecular level. Although in organisms that are not as sensitive as e.g.: insects, NNIs may cause geno-, immunotoxicity or oxidative stress, the mechanisms and long-term consequences are still unknown.

6 Conclusions

Neonicotinoids are clear stressors of natural ecosystems. They are highly mobile insecticides that can be detected in areas where they were not applied to and in organisms that were not the target of the application. Although a complete knowledge of their fate in the environment is missing, authorities are starting to react to the threat they pose by limiting their use and application. Relevant improvements have been made in the field of the toxicity to non-target organisms with the aim of protecting natural ecosystems and pollination processes. However, testing the toxicity of NNIs at more environmentally relevant conditions is a priority for the complete assessment of the threats to terrestrial and aquatic ecosystems. Studies that include factors such as mixture toxicity (Kunce et al. 2015; Maloney et al. 2017, 2018; Cavallaro et al. 2018; Rico et al. 2018), field or semi-field exposures (Beketov et al. 2008; Cavallaro et al. 2018; Pickford et al. 2018) and pulse exposures (Beketov et al. 2008; Beketov and Liess 2008; Mohr et al. 2012; Raby et al. 2018a) are of great value and need to motivate the performance of new experiments. They may contribute to renew the existing risk assessment data on single compounds in laboratory conditions by expanding the analyses to multiple compounds and stressors.

References

Addy-Orduna LM, Brodeur JC, Mateo R (2019) Oral acute toxicity of imidacloprid, thiamethoxam and clothianidin in eared doves: a contribution for the risk assessment of neonicotinoids in birds. Sci Total Environ 650:1216–1223

Alford A, Krupke CH (2017) Translocation of the neonicotinoid seed treatment clothianidin in maize. PLoS ONE 12(3):e0173836

Baxter, CV Fausch, KD Saunders, WC (2005) Tangled webs: reciprocal flows of invertebrate prey link streams and riparian zones. Freshwater Biol 50 (2):201–220

Beketov MA, Liess M (2008) Acute and delayed effects of the neonicotinoid insecticide thiacloprid on seven freshwater arthropods. Environ Toxicol Chem 27(2):461–470

Beketov MA, Schafer RB, Marwitz A, Paschke A, Liess M (2008) Long-term stream invertebrate community alterations induced by the insecticide thiacloprid: effect concentrations and recovery dynamics. Sci Total Environ 405(1–3):96–108

Bishop CA, Moran AJ, Toshack MC, Elle E, Maisonneuve F, Elliott JE (2018) Hummingbirds and bumble bees exposed to neonicotinoid and organophosphate insecticides in the Fraser Valley, British Columbia. Canada Environ Toxicol Chem 37(8):2143–2152

Bonmatin J-M, Giorio V, Girolami D, Goulson D, Kreutzweiser DP, Krupke C, Liess M, Long E, Marzaro M, Mitchell EAD, Noome DA, Simon-Delso N, Tapparo A (2015) Environmental fate and exposure: neonicotinoids and fipronil 22 (1): 35-67

Botías C, David A, Hill EM, Goulson D (2016) Contamination of wild plants near neonicotinoid seed-treated crops, and implications for non-target insects. Sci Total Environ 566:269–278

Brown M et al (2016) A horizon scan of future threats and opportunities for pollinators and pollination. PeerJ 4:e2249

Butchart SHM, Walpole M, Collen B et al (2010) Global biodiversity: indicators of recent declines. Science: 1187512

BVL, Federal Office of Consumer Protection and Food Safety, 2008 Background information: bee losses caused by insecticidal seed treatment in Germany in 2008. https://www.bvl.bund. de/EN/08_PresseInfothek_engl/01_Presse_und_Hintergrundinformationen/2008_07_15_hi_ Bienensterben_en.html Last accessed 30.10.2018

Bradford BZ, Huseth AS, Groves RL (2018) Widespread detections of neonicotinoid contaminants in central Wisconsin groundwater. PLoS ONE 13(10):e0201753

Byholm P, Mäkeläinen S, Santangeli A, Goulson D (2018) First evidence of neonicotinoid residues in a long-distance migratory raptor, the European honey buzzard (*Pernis apivorus*). Sci Total Environ 639:929–933

Cameron SA, Lozier JD, Strange JP, Koch JB, Cordes N, Solter LF, Griswold TL (2011) Patterns of widespread decline in North American bumblebees. PNAS 108(2):662–667

Canadian Council of Ministers of the Environment. 2007 Canadian water quality guidelines for the protection of aquatic life: imidacloprid. In: Canadian environmental quality guidelines, 1999, Canadian Council of Ministers of the Environment, Winnipeg

Cavallaro MC, Liber K, Headley JV, Peru KM, Morrissey CA (2018) Community-level and phenological responses of emerging aquatic insects exposed to 3 neonicotinoid insecticides: an in situ wetland limnocorral approach. Environ Toxicol Chem 37(9):2401–2412

Delfino Vieira CE, Perez MR, Acayaba RD, Montagner Raimundo CC, dos Reis Martinez CB (2018) DNA damage and oxidative stress induced by imidacloprid exposure in different tissues of the Neotropical fish *Prochilodus lineatus*. Chemosphere 195:125–134

European Food Safety Authority (EFSA) (2012) Statement on the findings in recent studies investigating sub-lethal effects in bees of some neonicotinoids in consideration of the uses currently authorised in Europe. EFSA J 10(6):2752

European Food Safety Authority (EFSA) (2013a) Conclusion on the peer review of the pesticide risk assessment for bees for the active substance clothianidin. EFSA J 11(1):3066

European Food Safety Authority (EFSA) (2013b) Conclusion on the peer review of the pesticide risk assessment for bees for the active substance thiamethoxam. EFSA J 11(1):3067

European Food Safety Authority (EFSA) (2013c) Conclusion on the peer review of the pesticide risk assessment for bees for the active substance imidacloprid. EFSA J 11(1):3068

European Food Safety Authority (EFSA) (2018a) Evaluation of the emergency authorisations granted by Member State Finland for plant protection products containing clothianidin or thiamethoxam. EFSA supporting publication 2018: EN-1419, 13 pp. https://doi.org/10.2903/sp. efsa.2018.en-1419

European Food Safety Authority (EFSA) (2018b) Evaluation of the emergency authorisations granted by Member State Romania for plant protection products containing clothianidin, imidacloprid or thiamethoxam. EFSA supporting publication 2018: EN-1416, 16 pp. https://doi.org/10.2903/sp.efsa.2018.en-1416

EFSA Panel on Plant Protection Products and their Residues (PPR) (2012) Scientific opinion on the science behind the development of a risk assessment of plant protection products on bees (Apis mellifera, Bombus spp. and solitary bees). EFSA J 10 (5):2668

Eng ML, Stutchbury BJM, Morrissey CA (2017) Imidacloprid and chlorpyrifos insecticides impair migratory ability in a seed-eating songbird. Sci Rep 7(1):15176

European Commission, Directorate General Joint Research Centre. Institute for Environment and Sustainability/H01-Water Resources Unit (2015) Development of the first watch list under the environmental quality standards directive. Publications Office of the European Union, Luxembourg, 166 pp. https://doi.org/10.2788/101376

Gallai N, Salles JM, Settele J et al (2009) Economic valuation of the vulnerability of world agriculture confronted with pollinator decline. Ecol Econom 68:810–821

Gibbons D, Morrissey C, Mineau P (2015) A review of the direct and indirect effects of neonicotinoids and fipronil on vertebrate wildlife. Environ Sci Pollut Res 22(1):103–118

Girolami V, Marzaro M, Vivan L, Mazzon L, Greatti M, Giorio C, Marton D, Taparro A (2012) Fatal powdering of bees in flight with particulates of neonicotinoids seed coating and humidity implication. J Appl Entomol 136:17–26

Godfray HCJ, Blacquiere T, Field LM, Hails RS, Petrokofsky G, Potts SG, Raine NE, Vanbergen AJ, McLean AR (2014) A restatement of the natural science evidence base concerning neonicotinoid insecticides and insect pollinators. Proc Royal Soc B 281:20140558

Goulson D (2013) Review: an overview of the environmental risks posed by neonicotinoid insecticides. J Appl Ecol 50(4):977–987

Grossman GM, Kruger AB (1995) Economic growth and the environment. Q J Econ 110 (2,1):353–377

Hallmann CA, Foppen RPB, Turnhout CAM, van Kroon HD, Jongejans E (2014) Declines in insectivorous birds are associated with high neonicotinoid concentrations. Nature 511:7509

Hladik ML, Kolpin DW, Kuivila KM (2014) Widespread occurrence of neonicotinoid insecticides in streams in a high corn and soybean producing region, USA. Environ Pollut 193:189–196

Hladik ML, Main AR, Goulson D (2018) Environmental risks and challenges associated with neonicotinoid insecticides. Environ Sci Technol 52(6):3329–3335

Hong X, Zhao X, Tian X, Li J, Zha J (2018) Changes of hematological and biochemical parameters revealed genotoxicity and immunotoxicity of neonicotinoids on Chinese rare minnows (Gobiocypris rarus). Environ Pollut 233:862–871

Hooper DU, Adair EC, Cardinale BJ et al (2012) A global synthesis reveals biodiversity loss as a major driver of ecosystem change. Nature 486:105–108

Huseth AS, Groves RL (2014) Environmental fate of soil applied neonicotinoid insecticides in an irrigated potato agroecosystem. PLoS ONE 9(5):e97081

Iturburu FG, Zoemisch M, Panzeri AM, Crupkin AC, Contardo-Jara V, Pflugmacher S, Menone ML (2017) Uptake, distribution in different tissues, and genotoxicity of imidacloprid in the freshwater fish Australoheros facetus. Environ Toxicol Chem 36(3):699–708

Iyaniwura TT (1991) Non-target and environmental hazards of pesticides. Rev Environ Health 9(3):161–176

Jeschke P, Nauen R (2008) Neonicotinoids—from ziro to hero in insecticide chemistry. Pest Manage Sci 64(11):1084–1098

Kunce W, Josefsson S, Örberg J, Johansson F (2015) Combination effects of pyrethroids and neonicotinoids on development and survival of Chironomus riparius. Ecotoxicol Environ Saf 122:426–431

Lee-Jenkins SSY, Robinson SA (2018) Effects of neonicotinoids on putative escape behavior of juvenile wood frogs (Lithobates sylvaticus) chronically exposed as tadpoles. Environ Toxicol Chem 9999:1–9

Lever JJ, van Nes EH, Scheffer M, Boscompte J (2014) The sudden collapse of the pollinator communities. Ecol Lett 17:350–359

Likens GE, Bormann FH (1974) Linkages between terrestrial and aquatic ecosystems. BioScience 24(8):447–456

Main AR, Michel NL, Cavallaro MC, Headley JV, Peru KM, Morrissey CA (2016) Snowmelt transport of neonicotinoid insecticides to Canadian Prairie wetlands. Agric Ecosyst Environ 215:76–84

Malev O, Klobučar RS, Fabbretti E, Trebše P (2012) Comparative toxicity of imidacloprid and its transformation product 6-chloronicotinic acid to non-target aquatic organisms: microalgae desmodesmus subspicatus and amphipod gammarus fossarum. Pestic Biochem Physiol 104(3):178–186

Maloney EM, Morrissey CA, Headley JV, Peru KM, Liber K (2018) Can chronic exposure to imidacloprid, clothianidin, and thiamethoxam mixtures exert greater than additive toxicity in *Chironomus dilutus*? Ecotoxicol Environ Saf 156:354–365

Maloney EM, Morrissey CA, Headley JV, Peru KM, Liber K (2017) Cumulative toxicity of neonicotinoid insecticide mixtures to *Chironomus dilutus* under acute exposure scenarios. Environ Toxicol Chem 36(11):3091–3101

Matsuda K, Buchingham SD, Kleier D, Rauh JJ, Grauso M, Satelle DB (2001) Neonicotinoids: insecticides acting on insect nicotinic acetylcholine receptors. Trends Pharmacol Sci 22(11):573–580

Mohr S, Berghahn R, Schmiediche R, Hübner V, Loth S, Feibicke M, Mailahn W, Wogram J (2012) Macroinvertebrate community response to repeated short-term pulses of the insecticide imidacloprid. Aquat Toxicol 110–111:25–36

Morrissey CA, Mineau P, Devries JH, Sanchez-Bayo F, Liess M, Cavallaro MC, Liber K (2015) Neonicotinoid contamination of global surface waters and associated risk to aquatic invertebrates: a review. Environ Int 74:291–303

Noriega JA, Hortal J, Azcarate FM et al (2018) Research trends in ecosystem services provided by insects. Basic Appl Ecol 26:8–23

Noyes PD, McElwee MK, Miller HD, Clark BW, Van Tiem LA, Walcott KC, Erwin KN, Levin ED (2009) The toxicology of climate change: environmental contaminants in a warming world. Environ Int 35(6):971–986

Pickford DB, Finnegan MC, Baxter LR, Bohmer W, Hanson ML, Stegger P, Hommen U, Hoekstra PF, Hamer M (2018) Response of the mayfly (*Cloeon dipterum*) to chronic exposure to thiamethoxam in outdoor mesocosms. Environ Toxicol Chem 37(4):1040–1050

Pisa LW, Amaral-Rogers V, Belzunces LP et al (2015) Effects of neonicotinoids and fipronil on non-target invertebrates. Environ Sci Pollut Res Int 22:68–102

Potts SG, Biesmeijer JC, Kremen C, Neumann P, Schweiger O, Kunin WE (2010) Global pollinator declines: trends, impacts and drivers. Trends Ecol Evol 25(6):345–353

Raby M, Nowierski M, Perlov D, Zhao X, Hao C, Poirier DG, Sibley PK (2018a) Acute toxicity of 6 neonicotinoid insecticides to freshwater invertebrates. Environ Toxicol Chem 37(5):1430–1445

Raby M, Zhao X, Hao C, Poirier DG, Sibley PK (2018b) Chronic toxicity of 6 neonicotinoid insecticides to Chironomus dilutus and Neocloeon triangulifer. Environ Toxicol Chem 37(10):2727–2739

Rico A, Arenas-Sanchez A, Pasqualini J, Garcia-Astillero A, Cherta L, Nozal L, Vighi M (2018) Effects of imidacloprid and a neonicotinoid mixture on aquatic invertebrate communities under Mediterranean conditions. Aquat Toxicol 204:130–143

RIVM (2014) Water quality standards for imidacloprid: proposal for an update according to the water framework directive. In: Smit CE (ed) National institute for public health and the environment. Bilthoven, Netherlands

Simon-Delso N, Amaral-Rogers V, Belzunces LP, Bonmatin JM, Chagnon M, Downs C, Furlan L, Gibbons DW et al (2015) Systemic insecticides (neonicotinoids and fipronil): trends, uses, mode of action and metabolites. Environ Sci Pollut Res 22 (1):5–34

Sur R, Stork A (2003) Uptake, translocation and metabolism of imidacloprid in plants. Bullet Insectol 56(1):35–40

Taliansky-Chamudis A, Gómez-Ramírez P, León-Ortega M, García-Fernández AJ (2017) Validation of a QuECheRS method for analysis of neonicotinoids in small volumes of blood and assessment of exposure in Eurasian eagle owl (*Bubo bubo*) nestlings. Sci Total Environ 595:93–100

Tan K, Chen W, Dong S, Liu X, Wang Y, Nieh JC (2014) Imidacloprid alters foraging and decreases bee avoidance of predators. PLoS ONE 9(7):e102725

Tomizawa M, Casida JE (2011) Neonicotinoid insecticides: highlights of a symposium on strategic molecular designs. J Agric Food Chem 59(7):2883–2886

Turaga U, Peper ST, Dunham NR, Kumar N, Kistler W, Almas S, Presley SM, Kendall RJ (2016) A survey of neonicotinoid use and potential exposure to northern bobwhite (*Colinus virginianus*) and scaled quail (*Callipepla squamata*) in the Rolling Plains of Texas and Oklahoma. Environ Toxicol Chem 35(6):1511–1515

United States Environmental Protection Agency (USEPA) (1992) Memorandum NTN 33893 Data Submissions for pending registration, DP Barcode #D182987. USEPA Archive Document

Velisek J, Stara A (2018) Effect of thiacloprid on early life stages of common carp (Cyprinus carpio). Chemosphere 194:481–487

Wood TJ, Goulson D (2017) The environmental risks of neonicotinoid pesticides: a review of the evidence post 2013. Environ Sci Pollut Res Int 24(21):17285–17325

Yu SJ (2015) Classification of insecticides. The toxicology and biochemistry of insects, 2nd edn. CRC Press, Boca Raton, Florida, USA, pp 31–102

RAGE Exacerbate Amyloid Beta (Aβ) Induced Alzheimer Pathology: A Systemic Overview

Firoz Akhter, Asma Akhter, Kavindra Kumar Kesari, Ruheena Javed, Janne Ruokolainen and Tapani Vuorinen

Abstract Alzheimer's disease (AD) is the most common irreversible, progressive brain disorder which causes problems with memory, thinking and behavior with the age. Alzheimer's is the sixth leading cause of death in the United States. Combination of genetic, environmental factors like; chemical radiations, toxicants and mutagens are the main causes for neurodegeneration. Including with these factors some other events can produce early stages of AD, known as early stage AD, and lead to the same eventual distinctive final pathways in the late stages. Such stages could be characterized by neuroinflammation, oxidative stress and neurodegeneration. Furthermore, advanced glycation end products (AGEs) exacerbate amyloid beta (Aβ) has shown enhanced neurotoxicity. Considering these factors, we reinvestigated the role of AGE–RAGE interaction in AD pathology. Accumulation of AGEs is a normal feature of aging, but it becomes impaied in AD. AGEs are prominent in amyloid plaques and neurofibrillary tangles. Several lines of evidences demonstrate that AGE-RAGE interactions are critical for disease pathogenesis and it is at least partially responsible for extensive oxidative stress, inflammation, and neurodegeneration. Therefore many in vitro, in vivo and clinical studies have been focused on AGE–RAGE inhibitors, although their undesirable side effects and solubility issues may limits the usage. Therefore, it is needed to develop a potential, effective and multi-targeted inhibitors in order to prevent AGE induced neurological disorder.

F. Akhter (✉) · A. Akhter
Department of Pharmacology and Toxicology, University of Kansas,
Lawrence, KS 66045, USA
e-mail: firozakhter86@gmail.com

K. K. Kesari · J. Ruokolainen
Department of Applied Physics, Aalto University,
02150 Espoo, Finland

R. Javed
Guangzhou Institute of Pediatrics, Guangzhou Women and Children's Medical Center,
Guangdong 510623, China

T. Vuorinen
Department of Bioproducts and Biosystems, Aalto University,
Espoo, Finland

© Springer Nature Switzerland AG 2019
K. K. Kesari (ed.), *Networking of Mutagens in Environmental Toxicology*, Environmental Science,
https://doi.org/10.1007/978-3-319-96511-6_9

1   Introduction

Alzheimer's disease (AD) is the foremost cause of dementia in the elderly population. Over 35 million people suffering with AD, has been reported worldwide in 2010 (Querfurth and LaFerla 2010). The World Alzheimer Report 2013 states that this number is expected to be double by 2030. AD is an advanced neurodegenerative disease that leads to the irreversible loss of neurons, intellectual abilities and eventually causes death within a few years (Akhter et al. 2017a, b, c). Major pathological hallmarks of AD include intracellular deposition of neurofibrillary tangles, which were associated with paired helical filaments (PHF), hyperphosphorylated tau protein and extra cellular accumulation of amyloid beta (Aβ) peptide in the senile plaques. Aβ is generated predominantly in a form of containing 42 amino acids from amyloid precursor protein (APP) on sequential cleavage by β and γ secretases (Yan et al. 2018). Indeed Aβ is a neurotoxic peptide and intra- and extra-cellular accumulation of this peptide was characterized by aggregation, insolubility, protease resistance and delayed turnover. This may leads to functional modifications of the cells including inflammation, impaired energy metabolism, overwhelmed oxidative stress, synaptic dysfunction followed by neuronal death (Akhter et al. 2017a, b, c). Another characteristic feature of AD is the intracellular accretion of neurofibrillary tangles (NFTs) in pyramidal neurons, which accentuates a local inflammatory state around the amyloid plaques and causes a decrease in glucose uptake and utilization by the brain cells (Schmidt et al. 2001). Despite extensive studies related to AD, and the exact pathophysiology of AD still remains elusive.

Several recent findings from different laboratories have reported that advanced glycation end products (AGEs) played a vital role in many disease settings including diabetes mellitus, inflammation, renal failure, atherosclerosis and neurodegenerative diseases, especially AD (Srikanth et al 2011; Ahmad et al. 2014; Galasko et al. 2014; Kuhla et al. 2014; Perrone and Grant 2015). These studies have shown that over accumulation of AGEs in brains of AD-affected people, is a major causative factor during an early stage of disease (Takeuchi et al. 2007). AGEs are the end products of post translational modifications of proteins, lipids and nucleic acids that should be facilitated by non-enzymatic reaction of reducing sugars and protein amino groups (Srikanth et al. 2009). AGE formation is accelerated under the conditions of oxidative stress, hypoxia, hyperglycemia and inflammation. The pathophysiological consequences of AGEs are mediated by reactive oxygen species (ROS), especially superoxide and hydrogen peroxide production. These pathological actions have also been mediated by cascade events of inflammation (Akhter et al. 2013, 2016, 2017a, b, c; Li et al. 2018; Shahab et al. 2014), which induces lowered glucose consumption, lowered ATP levels and down regulated mitochondrial activity that leads to the neuronal cell death (Kuhla et al. 2004). AGEs have been reported to impair the neuronal cells by cross linking with the substrates, resulting an increased Aβ deposition in AD (Yan et al. 2008).

Here we revisit the relation between advanced glycation end products and their receptor RAGE role in AD pathogenesis. Oxidative stress and neuronal inflammation caused by AGE-RAGE and/or RAGE-Aβ interactions are discussed and currently available inhibitors for AGEs and RAGE are documented.

2 AGEs and RAGE

Non-enzymatic reactions of sugar moieties and the amino group of proteins form AGEs and/or phospholipids (Ahmad et al. 2013, 2018; Shahab et al. 2014; Akhter et al. 2013). During AGE formation, Schiff bases undergo various reversible rearrangement steps to produce Amadori products. These products can rearrange again and eventually, through dehydration, condensation and oxidation reactions give irreversible compounds called as advanced glycation end (AGE) products. It is believed that two primary sources are responsible for AGEs are (1) food intake and (2) in vivo synthesis (Semba et al. 2014). Since the half-life of glycated products may double the normal cell life, this could to be a major part of their adverse effects on normal cells. However, long-lived cells such as nerve and brain cells, collagen and eye proteins may affected less. Nonetheless, the body as a whole is only able to combat the AGEs slowly. The pathological effects of AGEs is more concerned in the development of aging and age-related diseases including, AD (Deane et al. 2003; Chen et al. 2007), stroke (Muhammad et al. 2008), multiple sclerosis (Akhter et al. 2014), amyotrophic lateral sclerosis (Ilzecka 2009), neurological complications of diabetes (Toth et al. 2007), cancer, coronary artery disease and rheumatoid arthritis (Clynes et al. 2007; Logsdon et al. 2007). RAGE is a transmembrane receptor, belonging to the family of super immunoglobulins and it is located on chromosome 6 in the MHC class III region (Neeper et al. 1992; Schmidt et al. 1992). The human RAGE precursor has 404 amino acids with five kinds of protein domains such as three extra cellular immunoglobulin domains (V and C type domains), a single transmembrane domain and a short 43 amino acid C-terminal cytosolic tail. AGEs are the primary ligands for RAGE and other ligands such as HMGB1 (high mobility group box 1), S100 proteins (calgranulin family); prominently among these Aβ peptide, found to be central players in pathogenic mechanisms of AD (Yan et al. 1996, 2000). RAGE consists of different isoforms such as full-length RAGE (fRAGE), N-truncated RAGE (ligand deficient form) and soluble RAGE (sRAGE) that were generated by alternative splicing and proteolytic cleavage mechanisms (Kalea et al. 2011). Indeed, sRAGE lacks the transmembrane and cytosolic domain, and thus it is neither anchored in the cell membrane nor stimulated the signaling pathways including fRAGE (Huttunen et al. 1999). Two types of sRAGEs exist, endogenous sRAGE (esRAGE) and RAGE cleaved by ADAM-10 and MMP-9, which may act as a decoy receptor, stopping the interaction of mRAGE with its ligands (Zhang et al. 2008). Keeping in view the functional roles of RAGE as a core connector to many chronic diseases, several lines of candidate mechanisms has been proposed regarding RAGE actions in the disease progression. Interaction of AGEs and RAGE induces oxidative stress and initiates the inflammatory cascade system (NADPH oxidase and activation of different kinases including MAPK like ERK1/2, SAPK/JNK, p38; PI3 K/Akt; JAK/STAT and

GSK3ß). Finally, RAGE accelerates the downstream signals through transcription factors like NF-kß and SP1 and leads to the cellular perturbations (Motoyoshi et al. 2014; Yamagishi et al. 2015; Salahuddin et al. 2014; Shemirani and Yazdanparast 2014). Cellular dysfunctions are responsible for vascular damage, organ damage and/or failure. Earlier studies indicate that RAGE induces the influx of circulating Aβ from blood to brain through the Blood Brain Barrier (BBB). Further RAGE is antagonized by LRP-1 that mediates the efflux of Aβ (Sharma et al. 2012; Liang et al. 2013; Provias and Jeynes 2014; Wan et al. 2015). Up regulation of RAGE through neuroinflammation, tremendously increased Aβ levels in brain by enhancing the rate of Aβ influx, which is accompanied by decreased Aβ efflux (Kuwahara et al. 2013). On the other hand, the pathogenic mechanism of AD via RAGE-Aβ interaction followed the positive feedback mechanism driven by overwhelmed Aβ and maintained the RAGE expression (Akhter et al. 2017a, b, c). There has been growing interest in designing RAGE inhibitors and/or antagonists as a means to protect the AD affected brain from lipid peroxide derivatives and pathological consequences that result in Aβ removal (Thome et al. 1996). Recent observations by Zhang et al. (2014) have reported a genetic deficiency of neuronal RAGE that protects the synaptic damage via the AGE-mediated p38 MAP kinase system in knockout mice. The inability of available inhibitors to be effective here underscores the need to develop potent and safe drugs that can target multiple proteins to prevent this late life threaten.

3 RAGE and Amyloid Pathology

Many intriguing studies including biochemical, genetic and pathological studies revealed critical role of amyloid beta peptide (Aβ) in the AD pathogenesis (Fang et al. 2010). The amyloid hypothesis is well established in AD; it occurs due to imbalance in production and clearance of Aβ. The clearance of Aβ from the AD-affected brain indicates perivascular load of Aβ in the brain and RAGE plays a major role in the production of Aβ and the failure of its clearance.

3.1 Failure of RAGE in Aβ Clearance

It is well accepted from the published literature that amyloid hypothesis is a major considerable factor in AD pathogenesis. In an agreement to this, majority of the scientific interventions have focused on the development of therapeutics against AD through the Aβ clearance mechanism. Indeed the pathophysiological symptoms of AD were increased along with the Aβ accumulation due to failure in clearance mechanisms in the brain. The balanced circuit of Aβ production and removal will decide its role in physiology and/or pathophysiology of AD (Liu et al. 2012). The studies from different laboratories including our own have proposed different mechanisms for Aβ clearance such as microglial phagocytosis, interstitial fluid drainage and RAGE and lipoprotein receptor related protein-1 (LRP-1) mediated transport of Aβ across BBB to the circulation (Fang et al. 2010). It is well established that RAGE served as transporter of Aβ via influx and efflux actions. Since, efflux of Aβ from the brain into the circulation that could be employed by LRP-1 and P-glycoprotein. Further, vascular endothelial growth factor (VEGF) and endothelial nitric oxide synthase (eNOS) could also affect the Aβ transportation at the BBB. It has been suggested that endothelial network damaged by soluble Aβ accumulation and on the other hand RAGE mediated Aβ cytotoxicity to the vascular endothelial cells, eventually distort the BBB structural and functional integrity. RAGE-Aβ interactions would affect zonula occluden protein-1of tight junction, ultimately responsible for structural perturbations of BBB (Watson et al. 1991). Studies also reveals that Aβ mediates tight junction disruptions are harbored by RAGE via Ca^{+2} calcineurin pathway (Young et al. 2013; Wan et al. 2014). It is evidenced from the studies that lowered RAGE activity responsible for depletion of Aβ levels in cerebral blood vessels and mitigates the neuronal toxicity during AD.

3.2 Role of RAGE in Aβ Production

There are several studies documented that RAGE is the main factor for Aβ causing toxicity. The positive feedback mechanism of RAGE, when it interacts with Aβ, results in activation of the pro-inflammatory cytokines (IL-1 and IL-6) and ROS generation, which ultimately leads to further Aβ cytotoxicity, RAGE expression and BBB dysfunction. The interaction of Aβ oligomers mediates inflammation and oxidative stress, which aggravates AD pathogenesis. However, combination of Aβ monomer and RAGE did not show any effect (Qosa et al. 2014). AD-affected brain shows an increased expression level of RAGE that is possibly relevant to the Aβ generation through enhancing the activities of β and γ secreatases. Extensive recent studies indicated that oxidative stress is an event that precedes formation of neurofibrillary tangles (NFTs) and senile plaques, which are considered to be major hallmarks of AD. Therefore, RAGE is a crucial player in mediating oxidative stress and inflammatory response that could lead to a cycle of amyloid pathological change and neuronal stress followed by neuronal death.

4 Role of Oxidative Stress in AD

It is well established that, AGEs are involved in the vicious cycle of inflammation, generation of ROS that thereby initiates the copious production of AGEs and leading to an inflammatory environment. The interaction of RAGE-AGE may be linked to an increase in ROS production by numerous mechanisms, at least in part by causing perturbations in the numerous antioxidant defense systems (activities of superoxide dismutase, catalase and glutathione family related enzymes) and also activation of protein kinase C (Obrosova 2002; Jiang et al. 2004). The direct link between inflammation and AGE formation is indicated, where activation of myeloperoxidase (MPO) and its associated pathways directly generated CML-AGEs. Other studies have shown the oxidative stress feedback action of AGE, in which MDA (malonodialdehyde) cause secondary oxidative damage to the vital biomolecules including proteins (Traverso et al. 2004). Besides, activated microglia, mostly surrounded senile plaques are generating source of NO and O_2^- that can further react to form nitroxy radicals (Good et al. 1996). Aβ has been directly implicated in the formation of peptidyl radicals (Durany et al. 1999). Advanced glycation end products in the presence of transition metal ions (Cu, Fe ions) initiate redox cycling with consequent production of reactive oxygen and/or nitrogen species that could affect the cellular functions thereby accentuates the AD (Ramasamy et al. 2004).

5   AGEs-RAGE Interactions Provides Clues for Therapeutics for AD

The importance of AGEs-RAGE interaction was proposed in 1990s. In the human brain, AGEs are mainly located in pyramidal neurons (Sugaya et al. 1994) but over expressed in Alzheimer's patients. The rate of AGE-positive neurons increases along with the disease progression when compared to age-matched healthy subjects. Further, AGE-positive neurons exhibit hyperphosphorylated tau, suggesting a link between AGEs and NFTs (Luth et al. 2005). Choi et al. (2014) have reported that Aβ, AGE and RAGE positive granules were presented in the perikaryon of hippocampal neurons in AD. In AD brains most astrocytes (70–80%) contain both AGE and RAGE positive granules and were distributed in the major part of the brain, where as very few are contained with Aβ positive granules (20–30%) indicating the presence and distribution of glycated proteins other than Aβ. AGEs accelerate the Aβ deposition and plaque formation by providing driving force to Aβ polymerization (Loske et al. 2000). It is evident from the studies of Cho et al. (2009), who demonstrated that the AGE-RAGE interaction could regulate the amyloid precursor protein (APP) processing, which is considered to be an indicator for Aβ generation in neurons. In vitro cell cultures and AD transgenic animal studies reveal that AGE-RAGE interaction was responsible for increased dephosphorylation of the nuclear factor activated T-cells such as NFAT-1. In general, NFT1 is known to regulate the BACE1 expression at transcriptional levels. However, dephosphorylated NFAT1 by AGE-RAGE interaction could elevate the BACE1 expression, which subsequently enhances the Aß levels in the neuronal cells (Fig. 1a).

From the above information, it is clear that (1). AGE-RAGE interaction is critical for Aß generation; (2). NFT1 possibly regulate the BACE1 expression, which in turn regulates the APP processing; (3). Glycated Aß; It is hypothesized that these cohort approaches will be helpful to design potent drug molecules against AD.

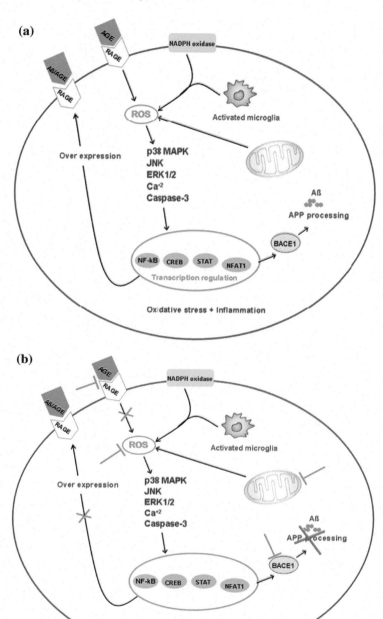

◄**Fig. 1** **a** The schematic figure represents AGE-RAGE interaction, aggravates the neuronal death in AD. AGE-RAGE interaction can initiates ROS generation; on the other hand NADPH oxidase and activated microglia could produce ROS. This ROS further activates the signaling pathways thereby it regulating Transcription; thus it forms a overwhelmed oxidative stress and inflammatory environment. Subsequently it accentuates the Aβ generation and reversible AGE/Aβ-RAGE complex, together have trigger neuronal death. **b** Keep in view of AD suffers health; there are several potential inhibitors have been experimentally tested against AGE-RAGE interactions, Aß and/or Aß-RAGE and different proteins of mitochondria. Since blockers can dwindle the ROS production by inhibiting the signaling cascade events subsequently it minimizes the oxidative and/or inflammation

6 AGEs-RAGE as Drug Targets

6.1 AGE Inhibitors

Several lines of evidence suggests that AGE-RAGE blockers are the promising candidates in AD therapy. Since, the blockers followed the different ways to inhibit the AGE formation, which includes (1) Blocking the attachment of sugar moiety to proteins, (2) Attenuation of glycoxidation (AGE cycle breakage) (3) Targeting of oxidative stress and/or inflammation. All targeting points depicted in the diagram (Fig. 1b).

Several synthetic compounds have been reported as AGE inhibitors. AGE inhibitor, Pimagedine (amino guanidine), which can deplete the AGE formation by interacting with the 3DG (Abdel and Bolton 2002). Deferoxamine, benfotiamine, carnosine and tenilsetam have also been reported as to down regulate the AGE activity (Rahbar and Figarola 2002). The reduction of Aβ toxicity may be achieved through the reduced inflammatory response and oxidative stress, which facilitate the detachment of AGEs to RAGE (Blatnik et al. 2008).

6.2 RAGE Inhibitors

Blatnik et al. (2008) have documented that mice treated with anti-RAGE antibody inhibit the binding of ligand to RAGE causing significant reduction in pro inflammatory cytokines indicating that RAGE can be used as a therapeutic target. In addition to this, other clinical trials have conclusively reported that RAGE antagonists such as TTP488 prevents the interaction of Aβ and RAGE thereby reducing the Aβ mediated cytotoxicity. On the other hand, soluble RAGEs have been tested clinically to block RAGE induced toxicity by competing with full length RAGE for ligand binding (Geroldi et al. 2006). Further, PF-04494700 clinically tested for efficacy for 10 weeks in AD-affected people did not reduce amyloid beta levels in plasma (Sabbagh et al. 2011). Recently, double blind, placebo-controlled clinical study at 40 academic centers in the US, PF-04494700 at 20 mg/day was shown to have adverse effects such as cognitive decline. It was, however, a good safety regimen at 5 mg/day concentration is under consideration for AD.

7 Conclusion and Future Perspectives

Advanced glycation end products (AGEs) were implicated in aging and age associated diseases including diabetes (type-II), atherosclerosis and neurodegerative disorders. In this way, the recent scientific interventions have led to reduce the AGEs and/or AGE mediated complications in resulting delayed the age onset diseases and improve the lifespan. RAGE plays a decisive role in the pathogenesis of AD, promoting the aging and related disorders by causing synaptic and neuronal circuit dysfunctions and Aβ-tau phosphorylation through the disrupted mitochondrial energy metabolism, oxidative stress and inflammation. We have previously reported that RAGE is a key cellular target that can be considered as potential drug target for AD (Yan et al. 2012). By taking, this basic principle in design and synthesis of different inhibitors for RAGE could be in use and under clinical trials. Since the available reports to date indicate there are no RAGE-inhibitors in clinical use for the AD patients, and there is an urgent need to develop a new generation of safe and durable molecules that inhibit RAGE activity, which can be used as potential drugs for clinical management of AD.

References

Abdel RE, Bolton WK (2002) Pimagedine: a novel therapy for diabetic nephropathy. Exp Opin Investig Drugs 4:565–574

Ahmad S, Akhter F, Moinuddin Shahab U et al (2013) Studies on glycation of human low density lipoprotein: a functional insight into physico chemical analysis. Int J Biol Macromol 62:167–171

Ahmad S, Akhter F, Shahab U, Rafi Z, Khan MS, Nabi R, Khan MS, Ahmad K, Ashraf JM, Moinuddin (2018) Do all roads lead to the Rome? The glycation perspective! Semin Cancer Biol 49:9–19. https://doi.org/10.1016/j.semcancer.2017.10.012

Ahmad S, Khan MS, Akhter F, Khan MS, Khan A, Ashraf JM, Pandey RP, Shahab U (2014) Glycobiology 24(11):979–90.

Akhter F, Akhter A, Ahmad S (2017a) Toxicity of protein and DNA-AGEs in neurodegenerative diseases (NDDs) with decisive approaches to stop the deadly consequences. https://doi.org/10.1007/978-3-319-46248-6_5

Akhter F, Chen D, Yan SF, Yan SD (2017b) Mitochondrial perturbation in Alzheimer's disease and diabetes. Prog Mol Biol Transl Sci 146:341–361

Akhter F et al (2013) Bio-physical characterization of ribose induced glycation: a mechanistic study on DNA perturbations. Int J Biol Macromol 58:206–210

Akhter F et al (2014) An immunohistochemical analysis to validate the rationale behind the enhanced immunogenicity of D-ribosylated low density lipo-protein. PLoS ONE 9(11):e113144

Akhter F et al (2016) Antigenic role of the adaptive immune response to d-ribose glycated LDL in diabetes, atherosclerosis and diabetes atherosclerotic patients. Life Sci 151:139–146

Akhter F et al (2017c) Detection of circulating auto-antibodies against ribosylated-LDL in diabetes patients. J Clin Lab Anal 31(2):e22039

Blatnik M, Frizzell N, Thorpe SR, Baynes JW (2008) Inactivation of glyceraldehyde-3- phosphate dehydrogenase by fumarate in diabetes—formation of S-(2-succinyl) cysteine, a novel chemical modification of protein and possible biomarker of mitochondrial stress. Diabetes 57:41–49

Chen X, Walker DG, Schmidt AM, Arancio O (2007) RAGE: a potential target for Aβ-mediated cellular perturbation in Alzheimer's disease. Curr Mol Med 7:735–742

Cho HJ, Son SM, Jin HS, Hong DH et al (2009) RAGE regulates BACE1 and Aβ generation via NFAT1 activation in Alzheimer's disease animal model. Faseb J 23:2639–2649

Choi BR, Cho WH, KimJ Lee HJ et al (2014) Increased expression of the receptor for advanced glycation end products in neurons and astrocytes in a triple transgenic mouse model of Alzheimer's disease. Exp Mol Med 46:75

Clynes R, Moser B, Yan SF, Ramasamy R (2007) Receptor for AGE (RAGE) weaving tangled webs within the inflammatory response. Curr Mol Med 7:743–751

Deane R, Yan S, Subramaryan RK, LaRue B et al (2003) RAGE mediates amyloid-beta peptide transport across the blood-brain barrier and accumulation in brain. Nat Med 9:907–913

Durany N, Munch G, Michel T, Riederer P (1999) Investigations on oxidative stress and therapeutical implications in dementia. Eur Arch Psychiatry Clin Neurosci 249:68–73

Fang B, Wang D, Huang M, Yu G, Li H (2010) Hypothesis on the relationship between the change in intracellular pH and the incidence of sporadic Alzheimer's disease or vascular dementia. Int J Neurosci 120(9):591–595

Fang F, Lue LF, Yan SQ et al (2010) RAGE-dependent signaling in microglia contributes to neuro inflammation, amyloid beta accumulation and impaired learning/memory in a mouse model of Alzheimer's disease. Faseb J 24:1043–1055

Galasko D, Bell J, Mancuso JY, Kupiec JW, Sabbagh MN, van Dyck C, Thomas RG, Aisen PS (2014) Alzheimer's disease cooperative study. Neurology 82(17):1536–1542

Geroldi D, Falcone C, Emanuele E (2006) Soluble receptor for advanced glycation end products: from disease marker to potential therapeutic target. Cur Med Chem 13:1971–1978

Good PF, Werner A, Hsu CW, Olanow DP (1996) Perl, evidence of neuronal oxidative damage in Alzheimer's disease. Am J Pathol 149:21–38

Huttunen HJ, Fages C, Rauvala H (1999) Receptor for advanced glycation end products (RAGE)-mediated neurite outgrowth and activation of NF-κB require the cytoplasmic domain of the receptor but different downstream signaling pathways. J Biol Chem 274:19919–19924

Ilzecka J (2009) Serum-soluble receptor for advanced glycation end products levels in patients with amyotrophic lateral sclerosis. Acta Neurol Scand 120:119–122

Jiang JM, Wang Z, Li DD (2004) Effects of AGEs on oxidation stress and antioxidation abilities in cultured astrocytes. Biomed Environ Sci 17:79–86

Kalea AZ, Schmidt AM, Hudson BI (2011) Alternative splicing of RAGE: roles in biology and disease. Front Biosci. 17:2756–2770

Kuhla A, Ludwig SC, Kuhla B, Münch G, Vollmar B (2014) Advanced glycation end products are mitogenic signals and trigger cell cycle reentry of neurons in Alzheimer's disease brain. Neurobiol Aging 36(2):753–761

Kuhla B, Loske C, de Arriba SG, Schinzel R (2004) Differential effects of advanced glycation end products and beta-amyloid peptide on glucose utilization and ATP levels in the neuronal cell line SH-SY5Y. J Neural Transmission 111:427–439

Kuwahara H, Nishida Y, Yokota T (2013) Blood-brain barrier and Alzheimer's disease. Brain and nerve Shinkei kenkyu no shinpo 65:145–151

Li Y, Khan MS, Akhter F, Husain FM, Ahmad S, Chen L (2018) The non-enzymatic glycation of LDL proteins results in biochemical alterations—a correlation study of Apo B100-AGE with obesity and rheumatoid arthritis. Int J Biol Macromol 1(122):195–200. https://doi.org/10.1016/j.ijbiomac.2018.09.107

Liang F, Jia J, Wang S, Qin W (2013) Decreased plasma levels of soluble low density lipoprotein receptor-related protein-1 (sLRP) and the soluble form of the receptor for advanced glycation end products (sRAGE) in the clinical diagnosis of Alzheimer's disease. J Clin Neuro Sci 20:357–361

Liu R, Wu CX, Zhou D, Yang F et al (2012) Pinocembrin protects against beta-amyloid-induced toxicity in neurons through inhibiting receptor for advanced glycation end products (RAGE)-independent signaling pathways and regulating mitochondrion-mediated apoptosis. BMC Med 10:105–116

Logsdon CD, Fuentes MK, Huang EH, Arumugam T (2007) RAGE and RAGE ligands in cancer. Curr Mol Med 7:777–789

Loske C, Gerdemann A, Schepl W, Wycislo M et al (2000) Transition metal-mediated glycoxidation accelerates cross-linking of beta-amyloid peptide. Eur J Biochem 267:4171–4178

Luth HJ, Ogunlade V, Kuhla B, Kientsch-Engel R et al (2005) Age- and stage-dependent accumulation of advanced glycation end products in intracellular deposits in normal and Alzheimer's disease brains. Cereb Cortex 15:211–220

Motoyoshi S, Yamamoto Y, Munesue S, Igawa H (2014) cAMP ameliorates inflammation by modulation of macrophage receptor for advanced glycation end-products. Biochem J 463:75–82

Muhammad S, Barakat W, Stoyanov S (2008) The HMGB1 receptor RAGE mediates ischemic brain damage. J Neurosci 28:12023–12031

Neeper M, Schmidt AM, Brett J, Yan SD (1992) Cloning and expression of a cell surface receptor for advanced glycosylation end products of proteins. J Biol Chem 267:14998–15004

Obrosova IG (2002) How does glucose generate oxidative stress in peripheral nerve? Int Rev Neurobiol 50:3–35

Perrone L, Grant WB (2015) Observational and ecological studies of dietary advanced glycation end products in national diets and Alzheimer's disease incidence and prevalence. J Alzheimers Dis 45(3):965–979

Provias J, Jeynes B (2014) The role of the blood-brain barrier in the pathogenesis of senile plaques in Alzheimer's disease. Int J Alzheimer's Dis 2014:1–7

Qosa H, LeVine H, Keller JN, Kaddoumi A (2014) Mixed oligomers and monomeric amyloid-beta disrupts endothelial cells integrity and reduces monomeric amyloid-beta transport across hCMEC/D3 cell line as an in vitro blood-brain barrier model. Biochim Biophys Acta 1842:1806–1815

Querfurth HW, LaFerla FM (2010) Alzheimer's disease. N Engl J Med 362 (4):329–344

Rahbar S, Figarola JL (2002) Inhibitors and breakers of advanced glycation end products. Curr Med Chem 2:135–161

Ramasamy R, Susan Vannucci J, Yan SD, Herold K (2004) Advanced glycation end products and RAGE: a common thread in aging, diabetes, neurodegeneration, and inflammation. Glycobiology 15:16R–28R

Sabbagh MN, Agro A, Bell J, Aisen PS et al (2011) PF-04494700, an oral inhibitor of receptor for advanced glycation end products (RAGE), in Alzheimer disease. Alzheimer Dis Assoc Disord 25:206–212

Salahuddin P, Rabbani G, Khan RH (2014) The role of advanced glycation end products in various types of neurodegenerative disease: a therapeutic approach. Cell Mol Biol Lett 19:407–437

Schmidt AM, Vianna M, Gerlach M, Brett J (1992) Isolation and characterization of two binding proteins for advanced glycosylation end products from bovine lung which are present on the endothelial cell surface. J Biol Chem 267:14987–14997

Schmidt ML, Zhukareva V, Perl DP, Sheridan SK, Schuck T, Lee VM-Y, Trojanowski JQ (2001) Spinal cord neurofibrillary pathology in Alzheimer disease and guam Parkinsonism-dementia complex. J Neuropathol Exp Neurol 60(11):1075–1086

Semba RD, Gebauer SK, Baer DJ, Sun K (2014) Dietary intake of advanced glycation end products did not affect endothelial function and inflammation in healthy adults in a randomized controlled trial. J Nutr 144:1037–1042

Shahab U et al (2014) Immunogenicity of DNA-advanced glycation end product fashioned through glyoxal and arginine in the presence of Fe^{3+}: its potential role in prompt recognition of diabetes mellitus auto-antibodies. Chem Biol Interact 219:229–240

Sharma HS, Castellani RJ, Smith MA, Sharma A (2012) The blood brain barrier in Alzheimer's disease: novel therapeutic targets and nano drug delivery. Int Rev Neurobiol 102:47–90

Shemirani F, Yazdanparast R (2014) The interplay between hyperglycemia—induced oxidative stress markers and the level of soluble receptor for advanced glycation end products (sRAGE) in K562 cells. Mol Cell Endocrinol 393:179–186

Srikanth V, Maczurek A, Phan T, Steele M et al. (2009) Advanced glycation end products and their receptor RAGE in Alzheimer's disease. Neurobiol Aging. https://doi.org/10.1016/j.neurobiolaging.2009.04.016

Srikanth V, Maczurek A, Phan T, Steele M et al (2011) Advanced glycation end products and their receptor RAGE in Alzheimer's disease. Neurobiol Aging 32:763–777

Sugaya K, Fukagawa T, Matsumoto K, Mita K et al (1994) Three genes in the human MHC class III region near the junction with the class II: gene for receptor of advanced glycosylation end products, PBX2 homeobox gene and a notch homolog, human counterpart of mouse mammary tumor gene int-3. Genomics 23:408–419

Takeuchi M, Sato T, Takino J, Kobayashi Y, Furuno S, Kikuchi S, Yamagishi S (2007) Diagnostic utility of serum or cerebrospinal fluid levels of toxic advanced glycation end-products (TAGE) in early detection of Alzheimer's disease. Med Hypotheses 69(6):1358–1366

Thome J, Kornhuber J, Munch G, Schinzel R et al (1996) New hypothesis on etiopathogenesis of Alzheimer syndrome. Advanced glycation end products (AGEs). Nervenarzt 67:924–929

Toth C, Martinez J, Zochodne DW (2007) RAGE, diabetes, and the nervous system. Curr Mol Med 7:766–776

Traverso N, Menini S, Maineri EP, Patriarca S et al (2004) Malondialdehyde, a lipoperoxidation—derived aldehyde, can bring about secondary oxidative damage to proteins. J Gerontol Biol A Sci Med Sci 59:890–895

Watson RR, Prabhala RH, Plezia PM, Alberts DS (1991) Effect of beta-carotene on lymphocyte subpopulations in elderly humans: evidence for a dose-response relationship. Am J Clin Nutr 53(1):90–94

Wan W, Chen H, Li Y (2014) The potential mechanisms of Aβ-receptor for advanced glycation end-products interaction disrupting tight junctions of the blood-brain barrier in Alzheimer's disease. Int J Neurosci 124:75–81

Wan W, Cao L, Liu L, Zhang C et al (2015) Abeta oligomer-induced leakage in an in vitro blood brain barrier model is associated with up-regulation of RAGE and metallo proteinases, and down-regulation of tight junction scaffold proteins. J Neurochem. https://doi.org/10.1111/jnc.13122

Yamagishi SI, Fukami K, Matsui T (2015) Crosstalk between advanced glycation end products (AGEs)-receptor RAGE axis and dipeptidyl peptidase-4-incretin system in diabetic vascular complications. Cardiovascular Diabetol, vol 14. https://doi.org/10.1186/s12933-015-0176-5

Yan JH, Rountree S, Massman P, Doody RS, Li H (2008) Alzheimer's disease and mild cognitive impairment deteriorate fine movement control. J Psychiatr Res 42(14):1203–1212

Yan SD, Chen D, Yan S, Guo L et al (2012) RAGE is a key cellular target for Aβ-induced perturbation in Alzheimer's disease. Front Biosci (School Ed) 212(2012):240–250

Yan SD, Chen X, Fu J, Chen M, Zhu H (1996) RAGE and amyloid-beta peptide neurotoxicity in Alzheimer's disease. Nature 382(1996):685–691

Yan SD, Zhu H, Zhu A, Golabek A (2000) Receptor-dependent cell stress and amyloid accumulation in systemic amyloidosis. Nat Med 6(6):643–651

Yan SF, Akhter F, Sosunov AA, Yan SD (2018) Identification and characterization of Amyloid-β accumulation in synaptic mitochondria. Methods in molecular biology, vol 1779. Clifton, N.J., pp 415–433. https://doi.org/10.1007/978-1-4939-7816-8_25

Young KS, Seok Hong H, Moon M, Mook-Jung I (2013) Disruption of blood brain barrier in Alzheimer disease pathogenesis. Tissue Barriers 1:e23993

Zhang H, Wang Y, Yan S, Du F et al. (2014) Yan, genetic deficiency of neuronal RAGE protects against AGE-induced synaptic injury. Cell Death Dis 5:e1288. https://doi.org/10.1038/cddis.2014.248

Zhang L, Bukulin M, Kojro E, Roth A (2008) Receptor for advanced glycation end products is subjected to protein ectodomain shedding by metallo proteinases. J Biol Chem 283:35507–35516

Elucidation of Scavenging Properties of Nanoparticles in the Prevention of Carcinogenicity Induced by Cigarette Smoke Carcinogens: An In Silico Study

Qazi Mohammad Sajid Jamal, Ali H. Alharbi, Mohtashim Lohani, Mughees Uddin Siddiqui, Varish Ahmad, Anupam Dhasmana, Mohammad Azam Ansari, Mohd. Haris Siddiqui and Kavindra Kumar Kesari

Abstract Nanotechnology, a science dealing with particles at nano scale, is currently used in many fields including environmental management and medicine for welfare of human being. The economic development and quality of life have been improved through nanotechnology. The Polycyclic aromatic hydrocarbons (PAHs)

Q. M. S. Jamal (✉) · A. H. Alharbi
Department of Health Informatics, College of Public Health and Health Informatics, Qassim University, Al Bukayriyah, Saudi Arabia
e-mail: sajqazi@gmail.com

M. Lohani
Research & Scientific Studies Unit, College of Nursing & Allied Health Sciences, Jazan University, Jazan, Saudi Arabia

M. U. Siddiqui
Department of Health Information Management, College of Applied Medical Sciences, Buraydah Colleges, Al Qassim, Saudi Arabia

V. Ahmad
Health Information Technology Department, Jeddah Community College, King Abdulaziz University, Jeddah, Saudi Arabia

Q. M. S. Jamal · A. Dhasmana
Novel Community Global Educational Foundation, Sydney, NSW 2770, Australia

A. Dhasmana
Himalayan School of Biosciences and Cancer Research Institute, Swami Rama Himalayan University, Dehradun, Uttarakhand, India

M. A. Ansari
Department of Epidemic Disease Research, Institutes of Research and Medical Consultation (IRMC), Imam Abdulrehman Bin Faisal University, Dammam 31441, Saudi Arabia

Mohd. H. Siddiqui
Department of Bioenginnering, Faculty of Engineering, Integral University, Lucknow, Uttar Pradesh, India

K. K. Kesari
Department of Applied Physics, Aalto University, Espoo, Finland

© Springer Nature Switzerland AG 2019
K. K. Kesari (ed.), *Networking of Mutagens in Environmental Toxicology*, Environmental Science,
https://doi.org/10.1007/978-3-319-96511-6_10

and other toxicants have higher affinity to scaveng by nanopartilces. The structural properties and surface chemistry of nanoparticles are the players, further, extremely high surface area to volume ratio results in multiple enhancement of many beneficial properties. Hence, we have followed a methodology to compare the binding efficiency of nanoparticles and cigarette smoke carcinogens with selected enzymes involved in DNA repair pathways. The molecular interactions have been accomplished using PatchDock server and interestingly got significant interacting results for our hypothesis. PatchDock results showed nanoparticles could be able to trap cigarette smoke carcinogens efficiently in the cellular system. The highest obtained binding efficiency between 4-(methylnitrosamino)-1-(3-pyridyl)-1-butanone (NNK) versus Single wall carbon nanotube (SWcNT) is 2632 score in contrast with NNK versus Human MDC1 BRCT T2067D in complex (PDB ID: 3K05) shows 2454 score, which means NNK could interact with SWcNT more efficiently than 3K05. Another part of the study shows that the highest binding efficiency 4-(methylnitrosamino)-1-(3-pyridyl)-1-butanol (NNAL) versus SWcNT = 2746 score and NNAL versus Titanium dioxide (TiO$_2$) Rutile = 2110 score in contract with NNAL versus Human Thymine DNA Glycosylase(PDB ID: 2RBA) shows 1696 score. It is also signified that NNAL interact with SWcNT and TiO$_2$ rutile more efficiently than 2RBA. The results clearly signifying that **SWcNT/TiO$_2$** are binding with NNK/NNAL more efficiently than biomolecules.

Keywords Cancer · Cigarette smoke · Nanoparticles · TiO$_2$ · In silico

1 Introduction

1.1 Environmental Chemical Causing Cancer

Human population is constantly exposed to many environmental chemical compounds and exprimentaly, they have been reported to cause cancer or mutation (Wogan et al. 2004).

The mutagenic and carcinogenic factors are commonly found in environmental air, water including soil which can affect exogenously and endogenously. The pathological and physiological activities such as production of some metabolic products may also cause certain changes in cellular activities which result into human cancer (Wogan et al. 2004). A food associated primary cancer of liver has been reported by aflatoxin. A causal association between contact to aflatoxin, a strongly cancer causing mycotoxin of dietary staples in Asia and Africa, and elevated risk for primary liver cancer has been reported through the application of well-valida (Kensler et al. 2011).This research has also been reported a striking synergistic interaction between hepatitis B virus infection a flat and oxinin elevating liver cancer.

Chuang et al. have reported the risk of cancer by the consumption of tobacco products (Chuang et al. 2010). It has also been reported well that tobacco carcinogens and their DNA adducts play a significant role to induce a specific tobacco mediated cancer like polycyclic aromatic hydrocarbons (PAHs) and NNK reported for lung cancer (Hecht 2012).

1.2 Prevention of Carcinogenesis by Using Nanoparticles as Scavenger

Nanotechnology, actually means the use of the substances at their nano scale, is expected to improve the economic development and quality of life globally. Understanding of metabolic processes of nanoparticles is a strong powerful force in the development of nanotechnology. The nanoparticles offers not only size-dependant physical properties but also offer beneficial optical and magnetic effects which have been used for a number of biological/medical applications, like as a fluorescent biological marker, for the gene and drug delivery, for the detection of proteins, pathogens, Probing of DNA structure, for the treatment of cancer by tumor destruction via heating (hyperthermia), in tissue engineering, for the purification of biological molecules and cells, in the contrast enhancement of MRI, and phagokinetic studies etc. (Kudr et al. 2017). The list of utilities of nanomaterials to biology or medicine is ever escalating. Recently, some of the nanoparticles have been used in soil remediation to remediated the high molecular weight PAHs from the contaminated soils (Karnchanasest and Santisukkasaem 2007). Amphiphilic polymer nanoparticles have also been used as nano-absorbent for pollutants in aqueous phase (Shim et al. 2007).

The hunting capacities of the nanoparticles for PAHs and other toxicants could probably be credited to their higher affinity towards the xenobiotics. The structural properties and surface chemistry of nanoparticles are the players, further, extremely high surface area to volume ratio results in multiple enhancements of such properties (Dhasmana et al. 2014).

1.3 Potential of Nanoparticles (TiO2) in Reduction of Harmful Compounds

TiO_2 is biological Inert but in ultrafine form and in high conc. TiO_2 causes the fibrosis in tissues which may lead the cancer (Chen et al. 2014). In 2006, the carcinogenic risk of TiO_2 reviewed by International Agency for Research on Cancer (IARC) and remarked that it is "possibly carcinogenic to humans" (Group 2B) based primarily on studies in rats indicating lung tumors (IARC 2006). However, epidemiology studies conducted in North America and Europe, on more than 40,000 workers in the titanium dioxide industry at manufacturing locations reported neither link with an amplified risk of lung cancer nor with any other adverse lung effects (Council 2013).

However, study conducted on inhalation exposures to TiO_2 in rats can result in lung tumors and lung effects (Bermudez et al. 2004). It is generally thought that the rat is exclusively sensitive to the effects of "lung overload", with the production of chronic lung inflammation and lung fibros which result into tumor formation but it was not observed in other species including humans (Warheit et al. 2016). The IARC conclusion was based on studies that involved rat "lung overload" effects. But in low and definite conc., Ultra Fine TiO_2 significantly reduced the harmful compounds from the cigarette smoke (Deng et al. 2011). A number of chemical compounds have been found in Cigarette smoke aerosol which are present in both vapour phase as well as particulate (Rodgman and Perfetti 2013). Some important compounds of cigarette smoke are tar, nicotine and water. The toxic nature and health risk have been observed with Tar, PAHs and tobacco-specific nitrosamines (TSNAs) (Lee et al. 2012).

Titanate Nano Tubes (TNT) and Titanate nanosheets (TNS) have also been synthesized and used to extract harmful compounds in CS (Deng et al. 2011). Thus, TNS and TNT were introduced into cigarette filter to reduce harmful compounds including nicotine, tar, hydrogen cyanide, ammonia, phenolic compounds and selected carbonyls. Interestingly, TNT exhibits highly efficient reduction capability for the most of the harmful compounds. This might be related to the intrinsic properties of TNT (Deng et al. 2011).

Hence, we have followed a methodology to analyse the binding efficiencies of nanoparticles and cigarette smoke carcinogens. The molecular interactions have been accomplished using PatchDock server and interestingly got significant binding results for advantageous contribution of our hypothesis in the field of carcinogens.

2 Materials and Methods

The minimum system requirement for the completion of computational study is as follows.

2.1 Supported Operating Systems

Discovery Studio Visualizer is supported on the following operating systems:

- Microsoft® Windows 7 Professional
- Red Hat® Enterprise Linux® 4.0, Updates 4-7
- Red Hat Enterprise Linux 5, Retail, Updates 1-2
- SUSE® Linux Enterprise 10 (SP2).

2.2 Processor and RAM Requirements

- Processor: An Intel-compatible ≥ 2 GHz is required.
- RAM: A minimum of 2 GB of memory for the visualizer.

2.3 Disk Space Requirements

A standard installation of Discovery Studio Visualizer requires 272 MB of disk space on Windows and 454 MB on Linux.

2.4 Software

- Accelrys discovery studio visualizer (*Designing of crystal structure, visualizing and manipulating protein and crystal 3D structures*) (Dassault Systemes, BIOVIA Corp., San Diego, CA, USA).
- PatchDock (Docking server).
- Open Babel (File converter) (O'Boyle et al. 2011).
- PyMol 3D structure visualizer (The PyMOL Molecular Graphics System, Version 2.0 Schrödinger, LLC.).
- An Internet Browser and valid internet connection.

2.5 Preparation of 3D Structures of Nanoparticles

After studying the anatase crystal structure, using Accelrys Discovery studio, we have designed the TiO_2 rutile (Fig. 1a), TiO_2 anatase (Fig. 2a), fullerene (Fig. 3c) and single wall carbon nanotube (SWcNT) (Fig. 3d) 3D crystal structure.

2.5.1 The 3D Structure of Cigarette Smoke Carcinogens

The chemical structures of carcinogens NNK (Fig. 2a) and NNAL (Fig. 2b) (Jamal et al. 2012, 2017) were drawn on Chemsketch (www.acdlabs.com) followed by generation of their PDB structures by (http://accelrys.com/products/discovery-studio/) Discovery Studio visualizer and their PDB structures were generated using link (http://www.molecular-networks.com) for tool CORINA. CharMM force field was application and optimized, subjected to single step minimization through smart minimize algorithm for 1000 steps at RMS gradient of 0.01 s (Brooks et al. 2009).

2.5.2 Preparation of 3D Structures of Proteins

From the our earlier study we have selected DNA repair enzymes 1CKJ (mammalian protein casein kinase I), 2O8B (DNA mismatch repair protein Msh2), 3K05 (human MDC1 BRCT T2067D in complex), 3GQC (human Rev1-DNA-dNTP ternary complex), 1Q2Z (the 3D solution structure of the C-terminal region of Ku86), 1T38 (human o6-alkylguanine-dna alkyltransferase) and 2RBA (Human Thymine DNA Glycosylase) and their crystal structures of DNA repair enzymes were downloaded from protein data bank (www.pdb.org) (Jamal et al. 2012, 2017). Further the selected enzymes were interacted with nanoparticles using PatchDock analysis.

2.6 Docking Studies Using PatchDock Server

We have performed comparative interaction analysis between nanoparticles and selected enzymes using PatchDock server (Fig. 3).

PatchDock server (http://bioinfo3d.cs.tau.ac.il/PatchDock/).

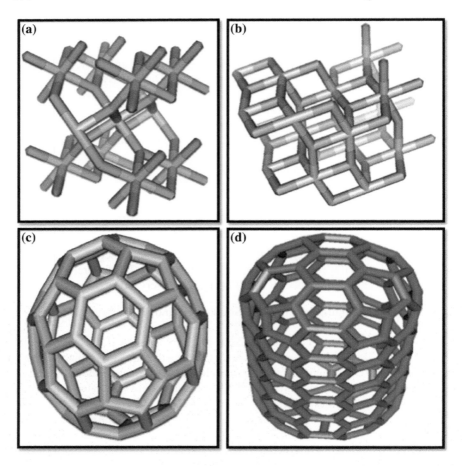

Fig. 1 Shows the structural variations of **a** TiO$_2$ rutile, **b** TiO$_2$ anatase, **c** fullerene and **d** SWcNT

- In the receptor molecule option click "choose file" button, and select the protein file "model.pdb", from the location where it has been saved. Then in the Ligand molecule option "choose file", and select the ligand file "Ligand.pdb", from the location where it has been saved.
- Give your e-mail address in the space provided where the results would be sent.
- Keeps the default clustering RMSD value , i.e., 4.0.

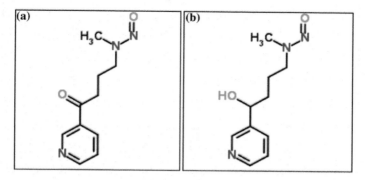

Fig. 2 Cigarette smoke carcinogens NNK and NNAL as a ligand for analysis, where structural compositions are **a** 4-(Methylnitrosamino)-1-(3-pyridyl)-1-butanone (PubChem Compound ID-47289, ChemSpider ID-43038), and **b** 4-(methylnitrosamino)-1-(3-pyridyl)-1-butan-1-ol (PubChem Compound ID-104856, ChemSpider ID-94646)

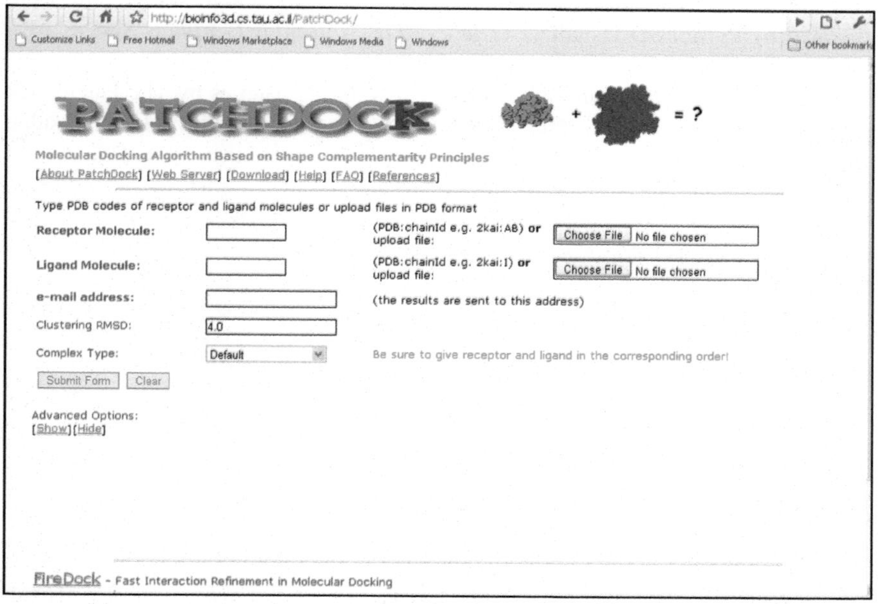

Fig. 3 Home page of PatchDock for the molecular docking algorithm

- Select complex type from the drop down menu as Protein-Small Ligand.
- Press "submit form" button. Results would be sent to the provided e-mail address after sometime.

3 Results and Discussion

We have performed molecular docking method using PatchDock server to find out the interaction between NNK versus Nanoparticles and NNK versus proteins involved in DNA repair Pathways (Table 1) and the interaction between NNAL versus Nanoparticles and NNAL versus proteins involved in DNA repair Pathways (Table 2).

The implemented hypothesis suggest that if NNK/NNAL and nanoparticles would be present in the cellular system than nanoparticles could interact with carcinogens like NNK and NNAL firstly on the basis of obtained binding energy using Patch-Dock tool and visualization of interaction pattern in Fig. 4a–u. All graphic were generated by PyMol 3D visualizer (*The PyMOL Molecular Graphics System, Version 2.0 Schrödinger, LLC.*).

The molecular surfaces are divided into shape-based patches by The PatchDock algorithm. This division deals the efficiency as well as discriminate between residue types (polar/non-polar) in the patches. Moreover, we also have created the use of

Table 1 Comparison of PatchDock scores obtained from docked NNK versus proteins and NNK versus nanoparticles conformations

S. No.	Protein's name	Protein versus NNK	SWcNT versus NNK	TiO$_2$ anatase versus NNK	TiO$_2$ rutile versus NNK	Fullerene versus NNK
1	1CKJ	2790	2632	2068	1360	910
2	2O8B	2720	2632	2068	1360	910
3	**3K05**	**2454**	**2632**	2068	1360	910
4	3GQC	3054	2632	2068	1360	910

Table 2 Comparison of PatchDock scores obtained from docked NNAL versus proteins and NNAL versus nanoparticles conformations

S. No.	Protein's name	Protein versus NNAL	SWcNT versus NNAL	TiO$_2$ anatase versus NNAL	TiO$_2$ rutile versus NNAL	Fullerene versus NNAL
1	1CKJ	3688	2746	2110	1360	954
2	1Q2Z	3374	2746	2110	1360	954
3	1T38	3240	2746	2110	1360	954
4	**2RBA**	**1696**	**2746**	**2110**	1360	954

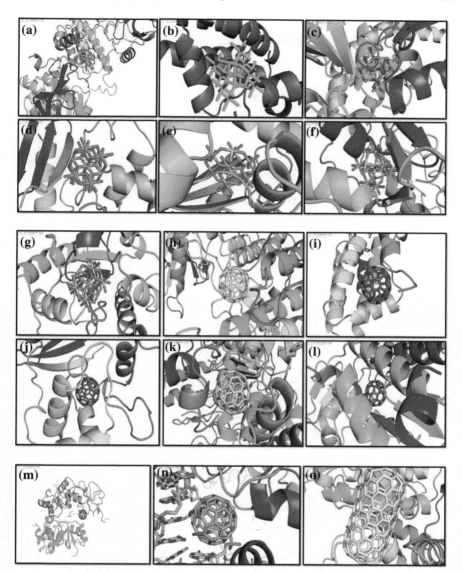

Fig. 4 **a** TiO$_2$ docked with 1CKJ; **b** TiO$_2$ docked with 1Q2Z; **c** TiO$_2$ docked with 2O8B; **d** TiO$_2$ docked with 1T38; **e** TiO$_2$ docked with 3GQC; **f** TiO$_2$ docked with 3K05; **g** TiO$_2$ docked with 2RBA; **h** fullerene docked with 1CKJ; **i** fullerene docked with 1Q2Z; **j** fullerene docked with 1T38; **k** fullerene docked with 2O8B; **l** fullerene docked with 3GQC; **m** fullerene docked with 3K05; **n** fullerene docked with 2RBA; **o** SWcNT docked with 1CKJ; **p** SWcNT docked with 1Q2Z; **q** SWcNT docked with 1T38; **r** SWcNT docked with 2O8B; **s** SWcNT docked with 3GQC; **t** SWcNT docked with 3K05; **u** SWcNT docked with 2RBA; **v** visualization of SWcNT and NNAL interaction; **w** visualization of SWcNT and NNK interaction

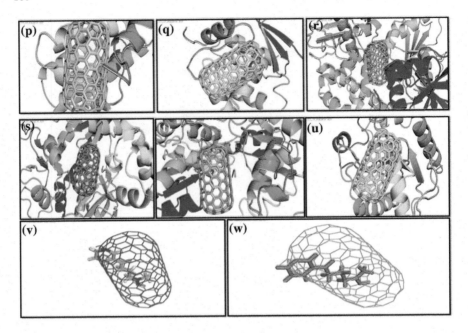

Fig. 4 (continued)

residue and hot spots in the patches. In the next step, improvement for the shape complementarily function are carried out by method by utilizing distance transform.

Moreover, it implements faster scoring, based on multi-resolution surface data structure. Our improved shape complementarily function further helps to improve the quality of the results. However, here the docking is rigid, the utilization of the last three components enables us to permit more liberal intermolecular penetration. PatchDock results showed nanoparticles could be able to trap cigarette smoke carcinogens efficiently in the cellular system. The highest obtained binding efficiency between NNK versus SWcNT is 2632 score (Table 1 and Fig. 4w) in contrast with NNK versus 3K05 shows 2454 score (Table 1), which means NNK could interact with SWcNT more efficiently than 3K05 (Fig. 4t). Another part of the study shows that the highest binding efficiency NNAL versus SWcNT = 2746 (Table 2 and Fig. 4v) score and NNAL versus TiO_2 Rutile = 2110 score (Table 2) in contract with NNAL versus 2RBA shows 1696 score (Table 2). It is also signified that NNAL interact with SWcNT and TiO_2 rutile more efficiently than 2RBA (Fig. 4g, u).

4   Conclusion

As mentioned earlier, the fact that technical TiO_2 has been very often of the (metastable) anatase form; and in many cases anatase is photocatalytically more active than rutile. This has provoked theoretical study of anatase, but there are hardly any experiments on well-characterized surfaces that would enable verification of these theoretical predictions. This lack of experimental research data is mostly because of the limited availability of anatase crystals of adequately large size.

This study of anatase [1, 0, 1] surface may establish valuable way for biotechnological researchers since this aspect of biotechnology has yet not been explored. The surface is also for interactions, this surface is also found over TiO_2 nanotubes, which are presently the subject of interest of the research community of electronics and nanotechnology innovators. In low and definite conc., TiO_2 significantly reduced the harmful compounds from the cigarette smoke (Deng et al. 2011).

The scavenging capacities of the nanoparticles for PAHs and other toxicants could probably be attributed to their higher affinity towards the xenobiotics. The structural properties and surface chemistry of nanoparticles are the players, further, extremely high surface area to volume ratio results in multiple enhancements of such properties.

Our study is conformity of study of Deng et al. (2011), who reported the use of titanate nanosheets and nanotubes are significantly reduces the harmful compounds in tobacco smoke. Our study confirmed this action in Biological system that by using of Bioinformatics tools we have done the comparative docking study between Nanoparticles-biomolecules and NNK/NNAL-Nanoparticles, we concluded that **SWcNT, TiO$_2$-Biomolecules** binding shown **lower** scores and **NNK/NNAL-Nanoparticles** binding shown **higher** scores. Hence, Results are clearly signifying that **SWcNT/TiO$_2$** binding with NNK/NNAL is more efficiently than biomolecules.

5   Future Scope

There is a lot to be done in this research work. We are just at the beginning; yet have much to be studied. Further studies can be done by applying force fields like crystal-CHARMm (Chemistry at HARvard Molecular Mechanics) (Brooks et al. 2009), SIBFA (Sum of Interactions Between Fragments Ab initio computed) (Gresh et al. 2002) (these are the few force fields which deals with metals and crystal structures) and then going for various interaction studies and energy calculations. The major hurdle in this work is that most of the softwares currently available do not recognize, i.e., they do not contain information regarding crystallographic bonding (or arrangements) and metallic atoms, their physical, chemical and quantum mechanical properties for molecular dynamic simulations of the same. There is an urgent requirement for a complete software package which can be used to design and manipulate inorganic or organic crystals as well as the biomolecules.

References

Bermudez E, Mangum JB, Wong BA, Asgharian B, Hext PM, Warheit DB, Everitt JI, Moss OR (2004) Pulmonary responses of mice, rats, and hamsters to subchronic inhalation of ultrafine titanium dioxide particles. Toxicol Sci 77:347–357

Brooks BR, Brooks CL, MacKerell AD, Nilsson L, Petrella RJ, Roux B, Karplus M (2009) CHARMM: the biomolecular simulation program. J Comput Chem 30(10):1545–1614

Chen T, Yan J, Li Y (2014) Genotoxicity of titanium dioxide nanoparticles. J Food Drug Anal 22(1):95–104

Chuang SC, Lee YCA, Hashibe M, Dai M, Zheng T, Boffetta P (2010) Interaction between cigarette smoking and HBV or HCV infection on the risk of liver cancer: a meta-analysis. Cancer Epidemiol Biomarkers Prev A Publication of the American Association for Cancer Research, Cosponsored by the American Society of Preventive Oncology 19(5):1261–1268. http://doi.org/10.1158/1055-9965.EPI-09-1297

Council TS (2013) About titanium dioxide. www.cefic.org/Documents/Industry%/20sectors/TDMA/About-TiO2-full-version-July-2013.pdf. 2013. Accessed 03 Dec 2015

Deng Q, Huang C, Xie W, Zhang J, Zhao Y, Hong Z, Pangb A, Wei M (2011) Significant reduction of harmful compounds in tobacco smoke by the use of titanate nanosheets and nanotubes. Chem Commun 47:6153–6155

Dhasmana A, Jamal QMS, Mir SS, Bhatt MLB, Rahman Q, Gupta R, Lohani M (2014) Titanium dioxide nanoparticles as guardian against environmental carcinogen benzo[alpha]pyrene. PLoS ONE 9(9):107068. https://doi.org/10.1371/journal.pone.0107068

Discovery Studio Visualizer (2017) Dassault Systemes, BIOVIA Corp., San Diego, CA, USA

Gresh N, Policar C, Giessner-Prettre C (2002) Modeling copper(I) complexes: SIBFA molecular mechanics versus ab initio energetics and geometrical arrangements. J Phys Chem 106(23):5660–5670

Hecht SS (2012) Lung carcinogenesis by tobacco smoke. Int J Cancer (Journal International Du Cancer) 131(12):2724–2732. http://doi.org/10.1002/ijc.27816

IARC (2006) Cobalt in hard metals and cobalt sulfate, gallium arsenide, indium phosphide and vanadium pentoxide. IARC Scientific Publications, vol 2006, p 86

Jamal QMS, Lohani M, Siddiqui MH, Haneef M, Gupta SK, Wadhwa G (2012) Molecular interaction analysis of cigarette smoke carcinogens NNK and NNAL with enzymes involved in DNA repair pathways: an in silico approach. Bioinformation 8(17):795–800

Jamal QMS, Lohani M, Dhasmana A, Siddiqui MU, Sayeed U, Wadhwa G, Siddiqui MH, Kesari KK (2017) Carcinogenic toxicity of cigarette smoke: a computational enzymatic interaction and DNA repair pathways. In: Kesari K (Ed) Perspectives in environmental toxicology. Springer International Publishing, Switzerland, pp 125–146. https://doi.org/10.1007/978-3-319-46248-6_6

Karnchanasest B, Santisukkasaem OA (2007) Preliminary study for removing phenanthrene and benzo[a]pyrene from soil by nanoparticles. J Appl Sci 7:3317–3321

Kensler TW, Roebuck BD, Wogan GN, Groopman JD (2011) Aflatoxin: a 50-year Odyssey of mechanistic and translational toxicology. Toxicol Sci 120(Suppl 1):S28–S48

Kudr J, Haddad Y, Richtera L, Heger Z, Cernak M, Adam V, Zitka O (2017) Magnetic nanoparticles: from design and synthesis to real world applications. Nanomaterials 7:243

Lee J, Taneja V, Vassallo R (2012) Cigarette smoking and inflammation: cellular and molecular mechanisms. J Dent Res 91(2):142–149

O'Boyle NM, Banck M, James CA, Morley C, Vandermeersch T, Hutchison GR (2011) Open babel: an open chemical toolbox. J Cheminform 3:33

Rodgman A, Perfetti TA (2013) The chemical components of tobacco and tobacco smoke (2nd edn), 1473. CRC Press, Taylor & Francis Group, Boca Taton, London, New York

Shim J, Park I, Kim J (2007) Use of amphiphilic polymer nanoparticles as a nano-absorbent for enhancing efficiency of Micelle-enhanced ultrafiltration process. J Ind Eng Chem 13(6):917–925

The PyMOL Molecular Graphics System, Version 2.0 Schrödinger, LLC

Warheit DB, Kreiling R, Levy LS (2016) Relevance of the rat lung tumor response to particle overload for human risk assessment—update and interpretation of new data since ILSI 2000. Toxicology 372:42–59

Wogan GN, Hecht SS, Felton JS, Conney AH, Loeb LA (2004) Environmental and chemical carcinogenesis. Semin Cancer Biol 14(6):473–486

Printed by Printforce, the Netherlands